DR PHILLIP GARTH LAW

HIS EXTRAORDINARY
LIFE AND TIMES

Don't be frightened to jump at an opportunity!

PHILLIP GARTH LAW (1912-)

To the memory of

Nellie Isabel 'Nel' Law

1914 ~ 1990

Dr Phillip Garth Law AC CBE

DR PHILLIP GARTH LAW

HIS EXTRAORDINARY
LIFE AND TIMES

as told to

IAN TOOHILL

THE ROYAL SOCIETIES OF AUSTRALIA
MELBOURNE - AUSTRALIA
2009

THE ROYAL SOCIETIES OF AUSTRALIA
Box 16 Church Street Brighton
Victoria 3186 Australia
rsa@scienceaustralia.org.au

ISBN 978-0-9806106-1-1 (hbk)
     978-0-9806106-0-4 (pbk)

National Library of Australia
Cataloguing-in-Publication data

    Law, P. G. (Phillip Garth), 1912-

        Dr Phillip Garth Law: his extraordinary life and times/ as told to
        Ian Toohill
        978-0-9806106-1-1 (hbk)
        9780980610604 (pbk)
        Notes: Taken from interviews conducted by Ian Toohill with
        Phillip and Nellie Law--Provided by publisher. Includes index.
        Subjects: Law, P. G. (Phillip Garth), 1912-
        Australia. Antarctic Division--History.
        Explorers--Australia--Biography.
        Scientists--Australia--Biography.
        Scientific expeditions--Antarctica--History.
        Antarctica-Discovery and exploration--Australian.
        Other Authors/Contributors:
        Law, Nellie Isabel, 1914-1990.
        Toohill, Ian, 1949-
    910.92

Proudly printed in Australia by
On-Demand Southbank Melbourne Victoria

CONTENTS

CONTENTS

ILLUSTRATIONS

MAPS

ACKNOWLEDGEMENTS

Camilla van Megen
Ian Toohill
Geoffrey Vaughan
Sandra Bishop
Wendy Coates
Julie Di Pietro
Bill McAuley

Australian Antarctic Division
Australian Science and Technology Heritage Centre
Janene Blanchfield Brown
Peter Law
Wendy Law Suart
Bronwyn Lowden
Jim Lowden
Helen Morgan
Basil Walby

Fred Elliott
David and Ellen Palmer Hubble
Photolibrary
Vladimir Sobolev
The Kelvin Club

DEDICATION

THE ROYAL Societies of Australia (RSA) in the case of The Royal Society of New South Wales, date from 1821. Their first national convention was held in 2004. Incorporation followed in 2007.

The RSA exists purely for **'the promotion and advancement of Science in Society for Society'.** Its Membership is restricted: open exclusively to the six, state-based Royal Societies of New South Wales, Tasmania, South Australia, Victoria, Queensland and Western Australia. H.E. The Governor General of The Commonwealth serves as Patron.

Australia's Royal Societies provide a joint history of nearly a thousand years with an immense reservoir of collective wisdom, influence and academic authority. The RSA aims to provide a fresh, egalitarian approach to the provision and co-ordination of science policy, education, outreach and communication, with its focus over coming years set squarely upon engaging Youth with both investigative and research Science in the national interest.

It is against this background that The RSA is very much indebted to **Dr PG Law AC CBE** (its first Chancellor) for providing it the honour of publishing Major Ian Toohill's collection of his interviews as an oral history.

A magnanimous and self-effacing man, PG Law has indeed led an extraordinary life throughout the last century. His insight, wit, intelligence and remarkable memory serve to capture the spirit and adventure of a bygone era: a time when a man's word was his bond: when capacity, duty, loyalty and perseverance meant something.

Many serve, but few are remembered. When I asked Dr Law to name the men he considered the greatest polar explorers, he ranked Sir Douglas Mawson 4th, Captain Robert Scott RN 3rd, Rear Admiral Richard Byrd USN 2nd and Sir Vivian Fuchs 1st.

To that company should be added the name Phillip Law: a man for all seasons; a seasoned man of destiny, who was arguably well-ahead of his time, the pack and his peers. He has managed to outlive them all and continues apace, from his residence in Melbourne, with the many things that interest him.

Long may his name and feats be remembered.

Dulce et periculum

WJW McAuley
President
The Royal Societies of Australia

26th January 2009
Australia Day

FOREWORD

I FIRST MET Dr Phillip Law in 1966 when he took up the position of Executive Vice-President of the Victoria Institute of Colleges (VIC). At the time I was the Deputy-Dean and Head of the Department of Chemistry at the Victorian College of Pharmacy. I was soon to be appointed to several of the committees of the VIC under the Chairmanship of Phil Law where, over time, I got to know him as a friend and colleague.

However I knew quite a bit about Phil before I had actually met him. I admired his contribution as an explorer and adventurer. From newspaper reports, like many Australians at the time, I knew that he was Director of Australia's Antarctic team and that he had established various stations in Antarctica; taken those long sea voyages in his beloved 'Dan' ships; and that he was married to Nel who had established her own profile as an Antarctic adventurer, and particularly as an artist.

As indicated from the above I thought I knew Phillip Law pretty well. But it has been through this book that I have learned much more.

Through his school-teacher father it is obvious that he had an early appreciation of discipline and education. As a country boy he developed an early love of sport, hiking, mountaineering and skiing which were the foundations he built on to become an adventurer and explorer. His education at school, teachers' college and university led to scientific qualifications in maths and physics which were to serve him well as an Antarctic explorer.

His achievement in gaining his degree through part-time study, and the transfer from country to city life, allowed him to show his tenacity. During this time Phil talks of "streaks of luck". I am certain that the reader, like me, will see this as determination; a characteristic which Phil showed throughout his life as a lecturer, scientist, explorer, government officer and educationalist. On top of this he showed exceptional leadership.

In all of his walks in life Phil Law was a superb organiser and administrator. From the time he became involved with Antarctica in 1947, through to retirement from the VIC in 1977, Phil was able to grow new and small organisations into viable and growing entities with a minimum of financial support, a minimum of bureaucratic help and the fight for government interest and support. Again this brilliantly shows the determination, tenacity and leadership of Phil Law.

The book gives a brief but wonderful account of Phil Law's association with Douglas Mawson; how Phil Law established Antarctic bases at Mawson, Davis and Casey; the challenges in arranging shipping, aircraft, landing craft and stores for the stations; and his own adventures and experiences during numerous voyages to carry out scientific observations and coastal and inland explorations. There is nothing but admiration one must have for his contribution to Australia's leadership position in Antarctic activities across exploration, meteorology, and many other areas of specialised Antarctic research.

It must be remembered that all this happened through the most trying physical conditions on earth through snow, ice, hurricanes and storms during 28 visits to Antarctica which included 11 major explorations. This was really something for a person who admits to suffer badly from seasickness. The seasickness was one of several reasons why Phil Law resigned as Director of the Antarctic Division in 1966 having devoted almost 20 years of working what seemed to be endless hours seven days a week. These reasons for retirement are outlined towards the end of the book.

Phil was snapped up by the Victorian Government in 1966 to establish the Victoria Institute of Colleges. This was because of his academic background, his administrative experience and his profile as a leader. Through this new role Phil had to bring together a number of non-university technical institutes and colleges to create a consortium of the newly titled Colleges of Advanced Education (CAEs). The VIC had to raise the academic standards of CAEs, improve their status, administrative structures and facilities, and where necessary provide new buildings, new campuses and even new colleges.

Phil Law, as was his experience in Antarctica, had to start from scratch, appoint a Council, a Board of Studies and a series of committees to advise on the requirements to create an Institute which would be recognised locally, nationally and internationally as a world leader in vocational education. He built the VIC as 'a pyramid of tertiary education, which was equivalent to, but different from the universities'.

In starting from scratch he had to personally find office premises and appoint his initial support staff single-handedly. His Antarctic experience stood him in good stead and again his determination and leadership shone through. He worked tirelessly to improve the lot of the CAEs and transformed higher education across Victoria, with many of his ideas and structures picked up by other administrations.

Amongst his many achievements across the VIC as a whole were the award of the first non-university degree in 1968, the Bachelor of Pharmacy at the Victorian College of Pharmacy and the first non-university post-graduate research degree in 1970, the Master of Pharmacy, again through the Victorian College of Pharmacy.

Other outstanding achievements include the establishment of the Victorian College of the Arts and the Lincoln Institute for Health Sciences. Phil retired from the VIC in 1976 after 11 years of outstanding leadership and achievement.

During his career and after retirement Phillip Law served as President, Chairman, Councillor or Committee Member of a wide range of academic institutions, scientific societies and community organisations, Through his lifetime he has been awarded honorary degrees, medals, and national and royal honours in recognition of his endeavours and for his outstanding contribution to a wide range of local, national and international activities.

This is the story of a truly remarkable Australian.

Geoffrey Vaughan AO

CHRONOLOGY

21 April 1912 Born Tallangatta, Victoria

1917-23   Student, Elsternwick State School, Gardenvale State School, Hamilton State School

1924-28   Student, Hamilton High School; senior swimming champion; member, School senior football and cricket teams

1929   Award of Merit, Royal Life Saving Society for swimming

             Junior teacher, Hamilton High School

1930   Junior teacher, Geelong High School

1931   Student, Ballarat Teachers' College; College colours for swimming, football and cricket

1932   First year science, University of Melbourne

             Matriculated, University of Melbourne

             Member, Intervarsity Boxing Team

             Novice and open lightweight boxing champion, University of Melbourne

             Secondary Teachers' Studentship (one year), Melbourne Teachers' College; College colours for swimming, boxing and football

1933-34   Teacher, Clunes Higher Elementary School

1935   Teacher, Coburg High School (short time)

1935-37   Teacher, Elwood Central School (mathematics, science and sports master)

1936   Open lightweight boxing champion, Intervarsity champion, University of Melbourne

1936-37	Member, Ski Club of Victoria
1938	Teacher, Melbourne Boys' High School (senior mathematics and physics)
1 April 1939	Graduated Bachelor of Science (BSc), First Class Honours in Physics, University of Melbourne
1939-40	Master's degree in physics, University of Melbourne (on leave without pay from the Education Department)
1940-41	Organised and led 'University National Service' student movement, University of Melbourne
1941	Enlisted and enrolled as a trainee Pilot Officer - Air Crew Navigator, Royal Australian Air Force, Point Cook, Victoria; Instructed to return to physics by the Commonwealth Manpower authority
5 April 1941	Graduated Master of Science (MSc), Honours, University of Melbourne
20 December 1941	Married Nellie Isabel 'Nel' Allan of Melbourne
1941-45	Researcher, Optical Munitions Panel (later Scientific Instruments and Optical Panel), Department of Munitions (under Professor Thomas Laby), University of Melbourne Assistant Secretary, Optical Munitions Panel
1941-47	Tutor of physics, Newman College, University of Melbourne
1943-48	Lecturer in physics, University of Melbourne

1943-48	President, Boxing Club, University of Melbourne
1944	One-man scientific mission to Papua New Guinea for the Australian Army
1945-65	Chairman, Australian National Antarctic Research Expeditions (ANARE) Planning Committee. This Committee comprised representatives of Commonwealth Government Departments, the Armed Services, the Australian Academy of Science and the universities.
1946-	Member, The Royal Society of Victoria
July 1947	Senior Scientific Officer, ANARE (seconded from the University of Melbourne)

December 1947 - March 1948

ANARE expedition to Antarctica and Macquarie Island in HMAS *Wyatt Earp* conducting experiments on latitude variation of cosmic rays

1947-75	Member, Alpine Club of Victoria

8 February - 1 April 1948

Senior Scientific Officer, Voyage in HMAS *Wyatt Earp* to Balleny Islands, Borradaile Islands, Commonwealth Bay and Macquarie Island

July-September 1948

Voyage to Japan in MV *Duntroon* by courtesy of the Australian Army to extend cosmic rays latitude variation measurements across the equator and tropics

1948	Fellow, Australian Institute of Physics
January 1949	Appointed leader, ANARE

21 January - 2 February 1949
Voyage in HMAS *Labuan* to Heard Island and Kerguelen Islands

18 July 1949 Fellow, Royal Geographical Society, London

1949 Member, Antarctic Club, London

1949-66 Annual relief voyages to resupply ANARE stations on Macquarie Island, Heard Island, and Mawson, Davis and Wilkes stations, Antarctica

Director, Antarctic Division, Commonwealth Department of External Affairs

Leader, ANARE

Personally led 23 voyages to Antarctica and sub-Antarctic regions, 11 of which explored the coast of the Australian Antarctic Territory from Oates Land in the east to Enderby Land in the west; directed ANARE activities both along the coast and inland that resulted in the mapping of 4000 miles of coast and 800,000 square miles of territory

January-March 1950
Australian observer, voyage in MV *Norsel* with Norwegian-British-Swedish Antarctic Expedition, Dronning (Queen) Maud Land

16 March 1950
Fellow, Institute of Physics

16 January - 1 March 1951
Voyage in HMAS *Labuan* to Heard Island and Kerguelen Islands

28 April - 19 May 1951
        Voyage in SS *River Fitzroy* to Macquarie
        Island

1951-52   Chairman, Victorian Division, Institute of
        Physics

9 February - 19 March 1952
        Voyage in MV *Tottan* to Heard Island and
        Kerguelen Islands

24 March - 16 April 1952
        Voyage in MV *Tottan* to Macquarie Island

1952-74   Chairman, Australian Committee on
        Antarctic Names

12-27 December 1953
        Voyage in MV *Kista Dan* to Macquarie
        Island

1953   Queen's Coronation Medal

1953-57   Member, Australian Committee,
        International Geophysical Year

4 January - 31 March 1954
        Voyage of MV *Kista Dan* to Heard Island,
        Kerguelen Islands and exploring in Mac
        Robertson Land

1954   Expedition to establish Mawson Station

7 January - 23 March 1955
        Voyage in MV *Kista Dan* to Heard Island
        Magnetic Island, Mawson and Kerguelen
        Islands

December 1955
        Appointment as Justice of the Peace,
        Australian Antarctic Territory

27 December 1955 - 26 March 1956
    Voyage in MV *Kista Dan* to Lewis Island, Balleny Islands, Mirny, Mawson, Heard Island and Kerguelen Islands

1955-56   President, Geographical Society of New South Wales

1956   Award of Merit, Commonwealth Professional Officers' Association

    Leader, Australian Delegation, International Geophysical Year Conference, Barcelona, Spain

17 December 1956 - 12 March 1957
    Voyage in MV *Kista Dan* to Vestfold Hills, Davis, Mawson and Kerguelen Islands

7-28 December 1957
    Voyage in MV *Thala Dan* to Macquarie Island

1957   Death of Lillie Lena Law

    Expedition to establish Davis Station

3 January - 19 March 1958
    Voyage in MV *Thala Dan* to Lewis Island, Dumont d'Urville, Mirny, Davis, Mawson, Heard Island and Kerguelen Islands

1958   Clive Lord Memorial Medal, Royal Society of Tasmania

6 January - 5 March 1959
    Voyage in MV *Magga Dan* to Lewis Island, Dumont d'Urville, Wilkes, Oates Land, Macquarie Island

1959   Expedition to take over Wilkes Station for Australia from the United States of America (now part of Casey Station)

1959-78 Member, Council, University of Melbourne

5 January - 11 March 1960

Voyage in MV *Magga Dan* to Dumont d'Urville, Davis, Wilkes, Lewis Island and Macquarie Island

29 November - 16 December 1960

Voyage in MV *Magga Dan* to Macquarie Island

1960 Founder's Gold Medal, Royal Geographical Society, London

Patron, British Schools' Exploring Society

22 December 1960 - 22 January 1961

Voyage in MV *Magga Dan* to Wilkes

24 January - 19 March 1961

Voyage in MV *Magga Dan* to Mawson, Dumont d'Urville, Oates Land and Macquarie Island

1-18 December 1961

Voyage in MV *Thala Dan* to Macquarie Island

1961 Commander of the Order of the British Empire (CBE), Civil division

22 December 1961 - 8 March 1962

Voyage in MV *Thala Dan* to Lewis Island, Wilkes, Commonwealth Bay, Dumont d'Urville, Oates Land and Macquarie Island

1962 Deputy-Chairman/ Chairman, World Health Organisation Conference on Medicine and Public Health in the Arctic and Antarctic, Geneva, Switzerland

1962   John Lewis Gold Medal, Royal Geographical Society of Australia (South Australia Branch)
Nella Dan launched

15 December 1962
    Doctor of Applied Science, *Honoris Causa*, University of Melbourne

9 January - 24 March 1963
    Voyage in MV *Nella Dan* to Heard Island, Mawson, Davis, Heard Island and Kerguelen Islands

1963-80   Member, Victorian State Committee, Duke of Edinburgh's Award Scheme

1963-88   Member, Recreation Grounds Committee, University of Melbourne

2-17 December 1964
    Voyage in MV *Nella Dan* to Macquarie Island

1964-66   Chairman, Working Group of the Committee of Investigation into Administration of the University of Melbourne; Member, Committee of Investigation

1964-74   Member, Council, LaTrobe University

1964-95   President, Geelong Area, Victorian Scouts Association

22 December 1964 - 15 March 1965
    Voyage in MV *Nella Dan* to Mawson and Davis for extensive exploration

29 December 1965 - 11 March 1966
    Voyage in MV *Nella Dan* to Wilkes, Mawson, Davis and Lewis Island

1966-77   Executive Vice-President, Victoria Institute of Colleges, responsible for co-ordinating the activities of 16 tertiary educational institutions

1966-80   Australian Delegate, Scientific Committee on Antarctic Research (SCAR), a committee of the International Council of Scientific Unions

1966-80   Chairman, Australian National Committee for Antarctic Research
Member, Council, Victorian Institute of Marine Sciences

1967-69   President, The Royal Society of Victoria

1968-77   Member, Advisory Council on Tertiary Education, Victoria

1968-78   Member, Board, Apex Foundation for Research into Mental Retardation

1968-83   Councillor, Science Museum of Victoria
Trustee, Science Museum of Victoria

November 1969
Polar Medal

1 March 1969 Two Thousand Men of Achievement Award

1969-70   President, Melbourne Film Festival

1969-72   Member, Victorian Universities and Schools Examinations Board

1969-77   Chairman, The Royal Society of Victoria Committee to establish an Institute of Marine Sciences

1970   Vocational Service Award, Rotary Club of Melbourne

1971-77	Member, Australian Council on Awards in Advanced Education
	Member, Australian Territories Accreditation Committee for Advanced Education
	President, Graduate Union, University of Melbourne
1972-77	Member, Committee for Natural Sciences, Australian National Committee for UNESCO
1972-77	Member, Victorian State Council for Technical Education
1972-79	Vice-President, Sports Union, University of Melbourne
1972-92	President, Melbourne Film Society
1972-	Trustee, Specific Learning Difficulties Association of Australia
1973	Death of Arthur James Law
	Men of Achievement Award
	Phillip Law Building constructed, Caulfield Institute of Technology (now part of Monash University)
1973-82	Member, Science and Industry Forum, Australian Academy of Science
1975	Foundation Fellow, Australian Academy of Technological Sciences and Engineering
	Doctor of Science, *Honoris Causa*, LaTrobe University
	Officer of the Order of Australia (AO), Civil division

1975-77	Member, Commonwealth Committee of Enquiry into Scientific and Technological Information Systems
1976-77	Chairman, Conference Organising Committee, Australian and New Zealand Association for the Advancement of Science (ANZAAS)
1976-79	Vice-President, Australian Club of Rome
1976-81	Chairman, Australian Committee on Antarctic Names
1976-82	Foundation President, Australian and New Zealand Scientific Exploration Society (ANZSES)
1977	Freeman, Victorian College of the Arts (VCA)
June 1977	Queen Elizabeth II Jubilee Medal
5 October 1977	Hon. Fellow, Royal Melbourne Institute of Technology (RMIT)
1977-80	President, Victorian Institute of Marine Sciences
1977-82	Member, Australia and New Zealand Schools Exploring Society
24 October 1978	Doctor of Education, *Honoris Causa*, Victoria Institute of Colleges
27 April 1978	Fellow, Australian Academy of Science
1978-80	Foundation President, Victorian Institute of Marine Sciences
1979	Member, Australian Delegation, Antarctic Treaty Consultative Meeting, Washington, USA

1979-82   Deputy President, Science Museum of Victoria

25 June 1981 Fellow, ANZAAS
               Member, Australian Delegation, Antarctic Treaty Consultative Meeting, Buenos Aires, Argentina

1982   Hon. Life Membership, Melbourne University Sports Union

August 1982 Sir Edmund Herring Memorial Award for Outstanding Service to the Youth of Victoria

1982-83   Chairman, Australian Scientific Exploration Foundation

1983   Member, Australian Delegation, Antarctic Treaty Consultative Meeting, Canberra, Australia

5 October 1983
               Patron, British Schools Exploring Society

1984-   Member, Antarctic Names and Medals Committee

10 November 1985
               Hon. Life Fellowship, Museum of Victoria

26 March 1985
               Statue of Victory Award, World Culture Prize for Letters, Arts and Sciences

12 March 1987
               Hon. Life Membership, Melbourne University Graduate Union

1987   James Cook Gold Medal, The Royal Society of New South Wales

1988	Gold Medal, Adventurer of the Year, Australian Geographic Society
29 April 1990	
	Death of Nellie Isabel 'Nel' Law
21 December 1994	
	Trustee, RMIT Foundation
1995	Companion of the Order of Australia (AC)
1995	Foundation Fellow, The Royal Society of Victoria
1995-98	Chairman, RMIT Foundation
2001	Member, Board of Directors, Glacier Society, USA
	National Award for Lifetime Contribution to Science and Technology, Clunies Ross Foundation
19 April 2002	
	Presented with bronze bust by Le Thanh Nhon (now in Phillip Law Building, Caulfield Campus, Monash University)
22 June 2002	
	Inaugural Annual Phillip Law Lecture, Antarctic Mid-Winter Festival, Hobart, Tasmania
2002	Portrait by Sally Robinson entered for the Archibald Prize
2004	The Dr Phillip Law Music Scholarship for music, VCA, University of Melbourne
2005	The Dr Phillip Law Travel Scholarship for dance, VCA, University of Melbourne
2006	Portrait by Ellen Palmer Hubble entered for the Archibald Prize in 2007

2008   Chancellor, The Royal Societies of Australia
      Doctor of Applied Science, *Honoris Causa*,
      RMIT

16 January 2009
      Phillip Law Room dedicated by HRH Prince
      Richard, Duke of Gloucester KG GCVO, The
      Royal Society of Victoria

INTERVIEWER'S NOTE

IT WAS about the year 1959 as a ten year old, that I watched a documentary about the British explorer Sir Vivian Fuchs making an epic journey to The South Pole during the International Geophysical Year (IGY) of 1957. I was mesmerised and decided then, that one day I would visit Antarctica. The opportunity came in 1982. I was a young Lieutenant in the Army Reserve based at the Sturt Street South Melbourne Army Transport (RACT) Depot, which was also home to the Army Detachment to ANARE. This detachment provided logistic support to Australia's Antarctic Program. I applied to my Commanding Officer, who supported my application, and after completing the requisite courses I was chosen as the third officer with the Army Detachment for the summer of 1982-83. So I went south as what was commonly known as a 'Larcie', thus fulfilling a life long ambition. Lighter Amphibious Resupply Cargo vessels (LARCs) took over The Army's ship to shore logistic support in 1970 from Army DUKWs and continued in that role until military support for ANARE ceased in 1993. The Australian Defence Force had been supporting ANARE for some forty-five years.

I was lucky to visit all three of Australia's continental Antarctic stations as well as Macquarie Island and spent the bulk of the summer at Mawson Station where I shot a documentary on life in Antarctica in between my military duties. This was a deal struck with the Education Department of Victoria which granted me leave from my role as a curriculum materials producer. The Army had no problem with this, as long as it did not impinge on my official duties. Consequently I worked very long hours under the midnight sun.

Returning to Australia in April 1983, I was keen to mount a photographic exhibition but needed someone to open the show at a gallery in Port Melbourne. My wife said, 'Why don't you give Phillip Law a call. I think he lives in Melbourne'. Dr Law graciously agreed and subsequently narrated my film 'Mawson Base Face to Face', which was produced by the Education Department of Victoria as a secondary school teaching resource.

I was very impressed with Dr Law's achievements and was determined to learn more of his life. Consequently in November 1984 I started a series of interviews with him, which concluded in June 2007.

My interviews, as transcribed by Sandra Bishop, are taken from original tapes. The text has received little editing and is, almost entirely, Phil and Nel Law's own words. There is some overlap and retelling of a few stories, but these have been left in to provide the slant as they stated it. This is after all an oral history.

The record of a life devoted to government and community service is indicated by the chronology, decorations and awards as well as the extensive bibliography. Dr Phillip Garth Law has clearly made an outstanding contribution to public life and science. It is therefore fitting that this oral history of the extraordinary life of a great Australian, be published at the end of the Fourth International Polar Year.

I feel both privileged and honoured to have been granted the opportunity to produce this book that has been twenty years in the making.

VICTORIA AND MELBOURNE

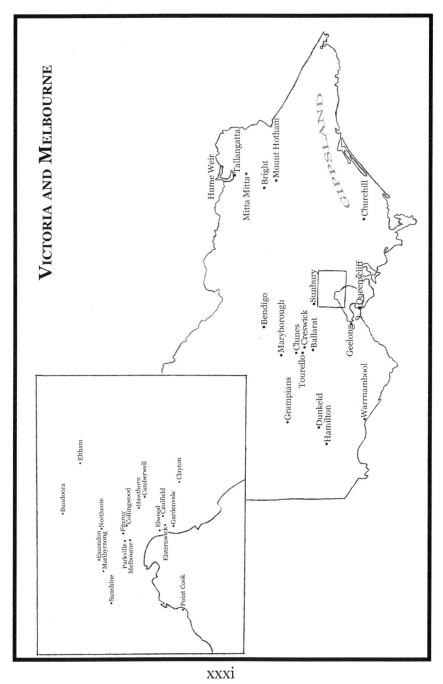

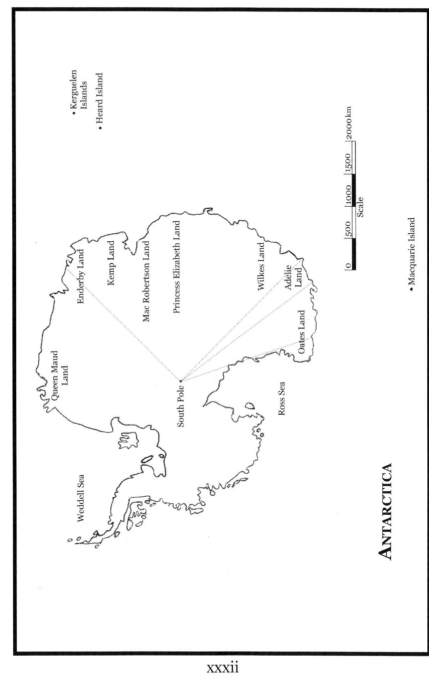

ANTARCTICA

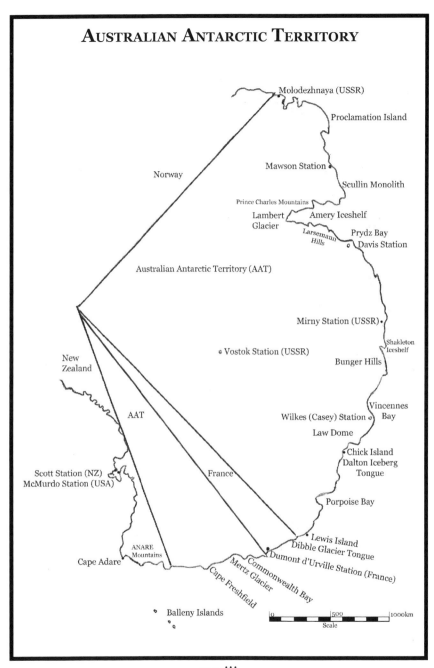

AUSTRALIAN ANTARCTIC TERRITORY

Molodezhnaya (USSR)

Proclamation Island

Norway

Mawson Station

Scullin Monolith

Prince Charles Mountains

Lambert Glacier

Amery Iceshelf

Larsemann Hills

Prydz Bay

Davis Station

Australian Antarctic Territory (AAT)

Mirny Station (USSR)

Shakleton Iceshelf

New Zealand

Vostok Station (USSR)

Bunger Hills

AAT

Vincennes Bay

Wilkes (Casey) Station

Law Dome

Chick Island

Dalton Iceberg Tongue

Scott Station (NZ)
McMurdo Station (USA)

France

Porpoise Bay

ANARE Mountains

Cape Adare

Lewis Island

Dibble Glacier Tongue

Dumont d'Urville Station (France)

Commonwealth Bay

Mertz Glacier

Cape Freshfield

Balleny Islands

0   500   1000km
Scale

xxxiii

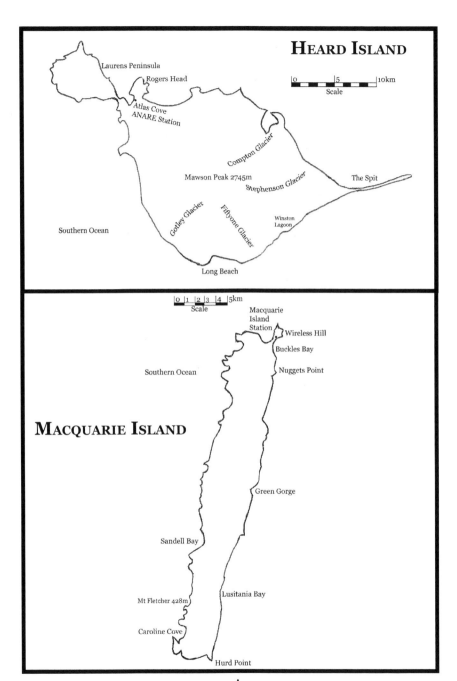

HEARD ISLAND

Laurens Peninsula

Rogers Head

Atlas Cove
ANARE Station

Compton Glacier

Mawson Peak 2745m

Stephenson Glacier

The Spit

Gotley Glacier

Fiftyone Glacier

Winston Lagoon

Southern Ocean

Long Beach

Scale
0   5   10km

MACQUARIE ISLAND

Scale
0 1 2 3 4 5km

Macquarie
Island
Station

Wireless Hill

Buckles Bay

Nuggets Point

Southern Ocean

Green Gorge

Sandell Bay

Lusitania Bay

Mt Fletcher 428m

Caroline Cove

Hurd Point

CHAPTER ONE

HIS EARLY YEARS

14 NOVEMBER 1984
THIS IS THE FIRST OF A NUMBER OF TAPES ON
THE LIFE OF DR PHILLIP LAW AC, WHO IS AN
ANTARCTIC EXPLORER OF CONSIDERABLE NOTE,
A FAMOUS AUSTRALIAN AND HAS HAD MANY
ROLES IN HIS LONG ACTIVE LIFE

Phillip could we perhaps start-off by you going right back to your early childhood and giving a little bit of background on where you were born, where you grew up and your school days?

Yes Ian, I was born in 1912 at Tallangatta, a little town in the north-east of Victoria, which disappeared beneath the waters of the Hume Weir a number of years ago. It reappeared as a new township on the banks of the weir, about five miles away from the site of the old flooded township.

My father [Arthur James Law] was Head Teacher of the little school at Mitta Mitta, which is about forty miles higher up the Mitta Valley from Tallangatta. My mother [Lillie Lena Law, nee Chapman] had to be driven forty miles, by gig and horse from Mitta to Tallangatta, to go to the hospital for my birth. I was the third of six children.

When I was about three and a half to four, my father was transferred to Melbourne. He was a lecturer at Melbourne

Teachers' College, which incidentally, he became principal of, many years later.

I had, up to that stage, lived at Tallangatta. My grandfather [James John Law] ran the Tallangatta newspaper which was called *The Upper Murray and Mitta Herald*. He ran that for over fifty years, so the 'Law' name was well known in Tallangatta.

I used to go back there for holidays as a young boy, because I had my grandfather there, as well as aunts and cousins. It was my favourite place and the hills there had a profound effect on my later interest in climbing, mountaineering and such activities.

When my brother and I were only six and eight years old, my grandfather would take us up the steep hill that was immediately behind the old township of Tallangatta. At any other place you would call it a mountain, but there were so many hills around that district, that this was only a hill, and has no particular name.

There is an interesting anecdote in regard to this. Having been introduced to this hill by my grandfather, I used to take great pleasure in climbing it every now and again. At the age of about eight or nine I climbed it. There's a long ridge of about six or eight miles that runs across and connects it to the highest mountain in that region, called Mount Charlie, which overlooks the higher Mitta Mitta River Valley. I'd always wanted to get to Mount Charlie and I remember on one occasion, setting out and climbing the hill behind Tallangatta, getting onto this long ridge, and hiking across it. Time ran out on me and I realised that I would get into trouble if I was late back. I didn't want to worry people, so I went back. It always stuck in my mind that I'd failed to get to Mount Charlie.

Now only last year, I was staying with some people at Tallangatta who live just across the river from Mount Charlie and their rural home looks straight at this mountain. After a good dinner that night, plus some nice red wine, we were talking and I mentioned how much I had always wanted to climb Mount Charlie. There it was and I'd still never climbed it. The lady of the house looked at me and said to me, 'Well, Phil, say we get up at five o'clock tomorrow morning and what say we climb Mount Charlie.' With the euphoria of the food and the grog I said, 'Yes, wonderful, tremendous!'

So I went to bed and as I was getting undressed I thought, what have I let myself in for because I haven't climbed anything for some years now. I didn't know whether I would run out of puff. Perhaps I'd get half way up and they'd have to send a helicopter to pick me up because I'd had a heart attack or something like that. So I thought she'd forget about that by the morning, she'd sleep in and that'd be the end of it. She was a young woman of about thirty-four, very fit and active. Her husband couldn't go because he had to go to work.

Sure enough, at five o'clock in the morning, she knocked on the door saying, 'Hey Phil, are you going to climb Mount Charlie?' So, I realised I was stuck with it and got up. We drove round the top of the weir, down the other side, arriving at the foot of Mount Charlie. She knew nothing about climbing, so I had to look at this and plan a route and, once done, we set off.

About two hours later we were on top of Mount Charlie, and at the age of seventy-two, this was very good for my ego. It was a straight up ascent, we went along the foot of it and then straight up. It's not a difficult climb, but

3

it is quite high, with quite a lot of puffing and panting, pushing through thistles in the lower places and through undergrowth higher up.

Do you have plans to go on and do some more?

Yes, well, I now want to go further up the Tallangatta Valley. There are some hills up the top of Valley, that which are very interesting. This lass wants to be in it too, because she's an active mountaineer type.

Well that time and date is perhaps indicative of the fact that you've always risen to challenges right throughout your life.

Well it's this interest in mountains, that I certainly got from the north-east and that was followed up later in life, when I lived in Hamilton, during my high school days. Hamilton is only twenty miles from the town of Dunkeld which nestles at the southern tip of the Grampians. My older brother was a very adventurous fellow, loving mountains also, so he and I started touring around in the Grampians.

What was his name?

Geof. He and I went on numerous trips. During almost every Easter, September and Christmas holidays that we had at school we'd be up in the Grampians. So that sort of hammered home the love of mountains that the Tallangatta experience had given me.

Then, from there, looking for fresh fields to conquer,

Lillie Lena Law (nee Chapman)
Wendy Law Suart collection

Arthur James Law
Wendy Law Suart collection

Arthur and Lillie Law
Wendy Law Suart collection

Phillip Law, aged 3
Phillip Law Collection

The three eldest Law children,
Phillip (left) with brother
Geoffrey and sister Marjorie,
about 1916
Wendy Law Suart collection

The three youngest Law
children, Peter, Dorothy,
mother Lillie and Wendy
Wendy Law Suart collection

Geoffrey and
his instruments;
Marjorie with
Dorothy; and
Phillip with his
bow and arrow
*Wendy Law Suart
collection*

Phillip, with Daisy air-rifle, and
Marjorie, 1924
Wendy Law Suart collection

Phillip trekking to the
Grampians with horse
Wendy Law Suart collection

Phillip (middle) at the Nerve Test in
the Grampians
Wendy Law Suart collection

Phillip Law, aged 13
Phillip Law collection

Phillip skiing with brother
Geoffrey at Mount Buffalo
Wendy Law Suart collection

Playing the accordion en route to
Mount Buller, about 1942
Phillip Law collection

Phillip (right) versus an
opponent from Sydney in his first
bout in the Lightweight
Australian University
Championship, 1936
Phillip Law collection

In New Guinea on
Optical Munitions
assignment, 1944
Phillip Law collection

we heard about the Australian Alps particularly around Mount Hotham, especially the fact that they had snow on them. So, while we were still at high school, my brother and I taught ourselves to ski and indulged in a number of exploits, getting into the whole alpine environment. This we both continued on with afterwards in our careers as teachers.

It was this skiing that led me to an interest in ice and snow. Then I learnt there was a place called Antarctica, and that there were books about Scott, Shackleton and Mawson. So by my mid-thirties, I was well up on Polar literature and all that sort of thing. Really, the direction of my life as regards Antarctic work, can be traced through the early development of a love of mountains, bushwalking and skiing.

Were there any particular experiences when you were in your academic school life, say secondary school for example, where you studied Antarctica or was it purely the mountains?

It was purely the bushwalking, mountaineering and skiing that took me there. Except that at Hamilton High School, I think, sport played a very important part. That turned me into at least as much a physical type, as an academic type. I was a fairly small boy.

How did you identify with that?

I was always about two years ahead of my age group in class, which made me appear even smaller than all the other boys in the class. That went on until intermediate level, when suddenly at the age of fourteen, I started to grow. Inside a couple of years I'd grown about six inches and so I was suddenly transformed from being a very much under-sized little whippersnapper, into a normal sized lad.

I was going to say that I was largely a bookworm, as a very small kid round about the end of primary school and into the lower grades at high school. I ranked number two in class, I remember, in the state school at Hamilton, where I just did one year in the state school. I was not dux; another bloke beat me for dux of the school.

I mention that because, he, and others in the top of that class went on to high school into form D. In those days it was first year high school, which is equivalent today to third year high school.

Oh I see because you went to eighth grade in primary school.

My father wouldn't let me go into form D. He said I was too young. He made me go into E form, which was a repeat of my merit certificate year at state school. I was pretty browned off because I was top of my class and here I was left a year behind all the others. That was bad enough, but at the end of my year in E form, the top half dozen in E form were allowed to jump D form and go straight into C form, which was the intermediate certificate year.

But not me, again my father, for a strange psychological reason and as a psychologist he should have known better, made me go into D form. Well, that finished me as a scholar. I decided 'to hell with them.' So I joined the sporting boys. I reckoned, I'm not going to do any more work, I'm not interested in a scholarship, I'll play sport. So in that year, in D form, I simply tagged onto the best sports boys in the school and worked like hell, trained like mad and tried to compete with them. I regard that as the turning point in my life in many ways because I suddenly became interested in sport.

I was very competitive and earnest and I trained like mad. I used to get up at six o'clock and run down to the swimming pool to train. I had some superb athletes as models, boys who I am eternally indebted to for setting standards for me to emulate.

Did any of them go onto achieve greatness?

No, none of them went on. One of them was the under fourteen athletic champion of Victoria, but he was a natural. He was a superb swimmer, footballer, tennis player, golfer and shooter with a gun or rifle. Chasing along behind him, I developed reasonable aptitudes in all sorts of sports. I believe that this accent on the physical part of my life had tremendous value later on, when I came to Antarctic things.

I was runner-up in tennis at school. I was never any good at athletics, although later, I took on long distance cross-country running. I played in the football and cricket teams, I was school champion swimmer and later on at university, I became Australian Lightweight University Boxing Champion. I got colours in cricket, football and boxing at Melbourne Teachers' College. I was never, except in boxing, an outstanding athlete, but I was reasonably good. I played sub-district cricket in Melbourne. I also played Victorian League thirds football in Melbourne and country football with a premiership team and so on.

Naturally, in Antarctic work, where an accent on physical fitness, toughness and competence was very important, it was also valuable in my leadership role in the sense of being able to compete with men on equal terms and lead by example.

Perhaps I had better go back a bit now. When we left Tallangatta we came down to Melbourne and lived in the suburb of Gardenvale which was just developing. I remember we lived near Kooyong Road, which was nearly a mile from the Gardenvale Station.

There were only about six houses between us and the Station, all the rest were empty blocks. I had to go to the Elsternwick State School, that was a mile and a half away, because there was nothing nearer. So that meant walking there every morning.

Was this in between going from Tallangatta to Hamilton?

Yes, so all my primary school years, except the last one in eighth grade, were spent at Gardenvale attending the Elsternwick State School. There was a toughening up process in the sense, as I used to have to walk this mile and a half to school every day and home again.

For a bit of a challenge, I used to sometimes see if I could run the whole way there without stopping. I don't think there was anything particularly notable about my time at Elsternwick State School, except the aspect of toughness that was there.

It was a pretty low grade school socially. In those days

it would be more of the level of Fitzroy or Collingwood. There were slums, as you know, in Richmond and Prahran stretching down that railway line and lots of the kids at that school were under-privileged.

It was a pretty rough and tough school. For example, there was a tradition of gang fights in the school. Different cliques would pick up sides and develop gangs and they'd have gang warfare. This led me to a certain amount of toughness and aggression and fights and fist fights and self protection and this sort of thing.

I remember the poverty of some of those kids. One kid used to come and perform antics hanging from branches of trees and emulating monkeys simply as a show at lunch times, so that he could then get sandwiches from other kids watching the performance. That was the only way that he got his lunch.

One adventurous part of this life was interesting. The Elwood Canal as you know runs down through Gardenvale and Elsternwick and down to the sea at Elwood. It wasn't as well developed then as it is now. It was more a creek with muddy banks rather than a concrete enclosed canal, but the drainage system was still the same.

While I was there as a kid, they built these great underground tunnel drains. They were all about six feet, or six foot six in diameter. They'd go for a mile underground and would pick up all the culvert discharged water from the streets and empty into the canal. We had great fun exploring these and we learnt them off perfectly. I could find my way round for miles underground, round the Elsternwick, Gardenvale area via these water conduits.

How did you get into them?

The normal way of getting into them would be to walk up the canal and go into the point where they disgorge. The other way would be at the end of the underground pipeline, there would be an iron ladder going up to a lifting lid and emptying into some deserted block. We'd know where they were so we'd get into the other end. But as a little kid I could also get in through the culverts.

I remember once being chased by one of these opposing gangs and tearing along and wondering where the hell to go. There was no way of escape and they were gaining on me, so I just dived down a gutter culvert in the middle of the street, much to the amazement of two old ladies who were walking along. They saw this shrimp of a boy disappear down a culvert and not come up again. That got me into a known part of this underground pipeline and I just went down to the canal that finished up near the school. I got out and went to school.

In my last year in Gardenvale they opened the Gardenvale State School. As a higher elementary school it had grades E and F, the first two grades in a high school, so I had about six months in the first form there before my father was transferred to Hamilton.

So at each stage in your schooling you went to the school where your father was Principal?

No, he was only Head Teacher at Mitta Mitta. When we were at Gardenvale he was a lecturer at Melbourne Teachers' College and when we went to Hamilton he was the District Inspector.

TEACHING SERVICE AND TERTIARY STUDIES

Well you obviously matriculated at Hamilton. What happened then?

In those days they had this system of junior teachers, so, having matriculated you had to do a year as a junior teacher before you went to teachers' college. I did my year as junior teacher at Hamilton High School, which was interesting again because as a junior teacher at a high school you weren't allowed to do any teaching, which is very different to being a junior teacher in a state school.

So, I was the private secretary to the principal and I had to learn to type. I typed all his correspondence and I was taught office work by him. He was a good organiser and a good administrator and I was taught such things as filing systems, typing and doing copies of things and so on.

My first exercise in administration was really under a very good headmaster. At this stage I was sixteen and a half, seventeen, something like that. At the end of that year, I was still not old enough to go into teachers' college because you had to be eighteen and I wasn't going to be eighteen until the following April.

My father was transferred to Geelong as District Inspector and I was able to transfer and become a junior teacher at Geelong High School. There again, it was very good experience, not in teaching, which I did none of, but it was a big high school and I was literally the secretary. I had to type all the exam papers and run them all off, do all the duplicating as well as being full time secretary to the headmaster. It was excellent experience.

At the end of that year, it was 1930, I had my sights fixed on getting into Melbourne Teachers' College, which was where my brother had been before me. I wanted to get in as a secondary student because that was the only way I could see of getting to university. But the Depression had struck and in 1931, when I was ready to go to teachers' college, they cancelled the secondary studentships and sent me to Ballarat. So instead of achieving my ambition of getting to university by going to Melbourne Teachers' College, I was sent to do a primary teachers' course at Ballarat Teachers' College, which was a one year course.

Well in one sense it was a great disappointment, but in another sense, it was again, extremely valuable. I found myself in a small college, only about sixty-six students, of whom about twenty-two were men. Social life was marvellous, because we had to create it ourselves. The sporting life was also marvellous because, again, we created it ourselves or moved into the district competitions in the Ballarat district.

I had a wonderful year of sport and scholarship, good lecturers, an excellent course at the Teachers' College and a broadening, in many ways, of my interests generally.

I found out that at the end of the year they picked two people, a woman and a man, to do what they called a second year. It was like a scholarship to go from there, down to Melbourne Teachers' College and start a first year of a university course.

I was lucky enough to get the position, so despite the disappointment of Ballarat in the first instance, it was a stepping stone that got me down to Melbourne Teachers' College, where I undertook first year science in 1932.

The scholarship was just a one year scholarship and my parents didn't have the means of supporting me any further, so after finishing first year science at Melbourne University, I had to go out and teach. That made things hard, because it meant that I couldn't go on with my degree work in the normal way. I was sent to Clunes Higher Elementary School, just outside Ballarat. This suited me as I knew the Ballarat area.

The two years, 1933 and 1934 there were probably the two most enjoyable years of my life. I became immersed in all the activities of a small country town. I was president of the cricket club and played in the tennis teams. I was a member of their swimming team that toured around and competed in Maryborough, Ballarat, Creswick and other places. I played football in the team that won the district premiership for the first time for thirty years.

All in all it was tremendous, but the only way that I could continue my degree was to proceed with mathematics because that was a subject where you didn't have to physically attend the university.

I got the syllabus and I borrowed some textbooks. A friend of mine who had done part one science with me the year before, was continuing on and I got him to take his notes from the maths lectures in duplicate and send me the carbon copies. So I did pure maths two and mixed maths two and pure maths three while I was doing those two years as a teacher.

Looking back on it now, I am absolutely staggered at having managed that because it was extremely difficult. Without those notes from my friend I could not possibly have done it. We would get practice examples to work out as exercises and I would just have to thrash through these and if I couldn't do them, I couldn't do them, there was no one to ask for help.

I think that the only thing that made me keep going and not throw it in was a fellow-teacher in a little rural school in Tourello, which is five miles outside of Clunes. He had decided to do mathematics for the same reason and every Friday night I'd walk five miles out across the paddocks to Tourello, to his place, and we'd spend Friday night until midnight swotting together on maths. Then I'd stay the night and Saturday morning he'd come into Clunes in his car to do the shopping and drive me home.

I had two years of working with this bloke and we pushed through maths two and three. If I hadn't passed those maths I would never have finished a science degree. I would have given it away.

What kept you going through those tough times with no support?

I don't know. Looking back on it I am staggered that I persisted because of all the temptations of the town life. The local lads were going to dances and things, while I would have to say 'No' and go back to my little room and swot. It was bitterly cold, I had a tiny, unheated bedroom, I'd wrap myself up in an overcoat and a rug and work there by oil lamp. There was no electric light.

There must have been some strong underlying motivation?

I suppose I had some sort of ambition to do something

scholastically. While on this question of scholastic attainment, I must follow through the rest of my university career because it didn't get any easier.

I might just mention one thing. While I was at Clunes, in my first year, I was doing pure maths part two and finding it very difficult and at the end of second term I decided I'd throw it in. However a few weeks before the exam my friend at Tourello, with whom I was studying, said 'Oh look Phil, you may as well have a go at the exam, it won't hurt, there's nothing to lose and everything to gain. We will get a day's holiday, go into Ballarat, do the exam, then go and have some fun and have a day off from work.' So on that premise, I accepted. But I'd chucked it, you see, in third term and I hadn't done any third term work, so I swotted up before the exam on my first two terms' work.

Then there was one of those lucky breaks you get in life. I went into the exam and there were certain alternatives, choices, and I found there were enough choices to allow me to complete the paper only on first and second terms' work. I was able to dodge all the questions on third term's work and I passed. There again, if I had failed that exam I would never have gone on with the degree.

Well, third year maths was interesting again. I was doing pure maths part three, and it contained a difficult new section that had just been introduced by Professor Cherry at Melbourne University. It comprised obtuse philosophical aspects of mathematics, which incidentally I found tremendously valuable in later life, but which, at that stage I found extremely difficult.

There was no precedence and no exam papers to look at, to see what sort of questions would be asked. When my friend and I went into this exam, we did as best as we could, but we were not very happy. We failed the first exam, but they gave us a supplementary, so that we could repeat it. I might mention that there were twelve people in total at the University of Melbourne who were doing pure maths part three and they only passed three out of twelve in that first exam.

Now that would be unheard of today, in a final year exam, because people who get as far as part three are generally reasonably good. They only passed three out of twelve, so I sat the 'supp' and I passed. Again there were only three of us who passed the 'supp'. That meant that out of twelve who sat for part three in pure maths that year, only six passed. I was lucky enough to be one of them.

Was that subject being offered at other universities?

Well there was only one university in Melbourne. It was interesting that my friend, who was studying full time at the university and who had sent me the notes in that second year, failed, and another friend failed. Each of them picked it up the following year but they had to repeat the year. So again I was extremely lucky just to sneak through.

Do you think luck really played a part or was it just sheer hard work and dedication?

Well it was a bit of each. I went down to Melbourne in 1935. I applied for a Melbourne school, mainly so I could go on with my university work. I wanted to do science. I had finished my mathematics part, but I had to pick up my science subjects. That meant practical work and you could only do that if you were in Melbourne. I was appointed for a few weeks to Northcote High School, but was transferred to Elwood Higher Elementary School.

I enquired at the University about part two science subjects. I already had part one physics and chemistry and I wanted to be a chemist, but there were no evening lectures in the chemistry school.

However there were evening lectures in the physics school and there was also an evening practice class for practical work in the physics school. So I was able to do part two physics by only attending lectures and prac work during the evenings and on Saturday mornings.

That got me through part two physics while I was teaching full time. I didn't get any time off. This swung me from an interest in chemistry to an interest in physics. My third year physics was completed during my second year at Elwood Higher Elementary School.

I was carrying a full load of teaching and I was also sportsmaster. In third year I found they had no evening lectures and no evening prac work. I found that I could attend lectures three times a week, between twelve and one o'clock, by arranging with the school to get off the last period which was between twelve and twelve-thirty. Twelve-thirty to one-thirty was lunch. So I'd get in my little car at a quarter to twelve, race up to the University, attend twelve to one, eat sandwiches in the car as I drove back to school and just get back in time to start afternoon teaching. So that year, my second year at Elwood, I did the course for physics part three.

At the end of the year, they told me I couldn't present for the exam because I hadn't fulfilled enough hours of practical work. You had to do between six and nine hours a week and I only got to do three hours on a weekend and three hours on Wednesday afternoons.

This was absurd, because I'd finished all the prac work. In those days they set you a certain number of experiments and I had to do twenty-two experiments. I had done them all, but had not complied with the number of set hours.

Then I had another one of those strokes of luck that have followed me all through my life. About, September or October the poliomyelitis epidemic swept Melbourne.

It was 1936, and our school was closed for five weeks. In those five weeks, I went into Melbourne University and visited the lab. I didn't do any prac work because I didn't have anything left to do, but I signed the book and got my hours up.

I spent the time swotting in the university library. That enabled me to get permission to sit the exam. With the extra swot, I was able to sit for honours. It was the first time I had sat for honours, as I had never had the confidence before with the part-time work. You had to do two papers to pass, and an extra paper for the honours. This time I thought I'd have a go and I was lucky enough to get First Class Honours. This gave me an entrée to postgraduate work. If I hadn't got honours I couldn't have done the postgraduate degree.

To finish off I'll just say that I did another year at Elwood. Then I went and had a year teaching at Melbourne Boys' High School.

At that stage, at the end of that year, I found that I was not getting anywhere in the teaching profession because of the complete deadlock on promotion. I was number three hundred and something on the promotions list and they were only promoting about ten or fifteen a year.

I could see that although I had a First Class Honours degree in Physics and an A class teaching mark (I couldn't have done any better in my profession.), it looked like being ten years before I got out of the bottom class. The bottom class was class five and I was still in it.

So I thought to hell with this, I'll get a couple of years leave from the Education Department and do a Master's degree. So I sold my little car and went into lodgings in Parkville, and I spent 1939 and 1940 doing a Master's degree, playing in a jazz band and doing a bit of tutoring for extra money. I finished my Master's by the end of 1940 and then, as I was ready to go back to teaching, of course the war had started.

I took out my Master's degree in March 1941 and I immediately enlisted in the RAAF [Royal Australian Air Force] as a navigation officer. I was accepted and put through all the tests and duly told to report to Point Cook to begin a course. I had received approval from my professor to do this.

I spent one day at Point Cook and at the end of the first day, the Commanding Officer summoned me into his presence. He said, 'Law, I am very sorry but you will have to go back to the university. Your professor has complained that he's losing staff and as he's involved in Optical Munitions work, he has invoked the Manpower Regulations!'

So I was returned to the University. That was highly embarrassing because, my friends there had given me various send-offs. I had also relinquished my various tutoring jobs, which was helping me earn money for the course. So i crawled back and suddenly appeared again when everyone had thought I'd gone off to war. Well that was the end, at that stage, of my academic career.

The Musician and Sportsman

Just going back a bit, you mentioned that you played in a jazz band for a couple of years to provide money to support yourself while you weren't working for the Education Department.

Oh yes, that was interesting. That started back at Hamilton when I was in the lower grades at high school. I became friendly with a bloke who was a groom in a big home in Hamilton. He was a very keen musician and he had a little room at the back of the stables. He had a variety of instruments and some of us used to go and listen to him and fiddle around with his instruments. He lent me a piccolo.

Although I'd taught myself the mouth organ and tin whistle, the piccolo was a step forward because it had sharps and flats, which a tin whistle hasn't got. So I taught myself the piccolo.

Then a chap in Hamilton had a broken-down clarinet, which he gave me. It had no springs and no pads, so I innovated. I got some blue hat pins, which were hardened steel, I cut them with pliers and used them as springs and I got some old kid gloves of my mother's and with these, cardboard, wadding and some shellac, I made the pads. I repadded this old clarinet and having repadded it and put in the springs, I had a clarinet. So I taught myself clarinet all by ear.

My elder brother, Geof, who was a wonderful musician, all self-taught, could play about five instruments. He set up the first state high school band I think in Victoria. It was not done with the help of the teachers or encouragement of anybody, it was a pure student initiative and he set up a five-man jazz band. I played drums in it because I wasn't good enough at clarinet at that stage.

I remember getting browned off playing drums because at the end of the dance everyone would leave and you'd be left to pack up all this gear in time to get it into a car and get home somewhere. I used to long to be a pianist where you'd just walk in, and the piano was there.

We were the number two jazz band in Hamilton which meant that if the first band was engaged and someone wanted a second dance in the town, we had to be employed. We got very good professional experience, playing around Hamilton, as high school students playing for money.

Then when I went to Ballarat Teachers' College, I organised a little jazz band there and when I went to Clunes, I teamed up with a man who's still a very close friend of mine, Theo Harden, who was a good pianist. He and I organised a little jazz band in Clunes.

We played at all the local dances, the usual Saturday night dance, mid-week dances and now and then at a big ball on a Wednesday or Thursday night. We played all around the district and in Ballarat occasionally.

That's when I really learnt to play, as I switched from clarinet to saxophone and taught myself to sight-read.

At Clunes, I also took my first primitive steps to teaching myself piano. So I look back on the Clunes days as being very valuable musically, as well as the work I told you about in my degree. In other words I had two very fulfilling years there.

How did you fit it all in?

All the sport? I used to go shooting, three or four nights a week after school, rabbits, foxes, ducks and quail. I developed quite a high ability in shooting, which I found extremely enjoyable in later years.

So with all that relaxation, with music and shooting you must have had very little time to sleep.

It was a tremendous couple of years, yes. Well when I came down to Melbourne, I kept my jazz music going. I teamed up with a man called Ernest 'Dick' Gardiner. Now Dick Gardiner later became Senior Master at Melbourne Grammar School. He became a very important person in curriculum design and policy for the State Government and later for the Commonwealth Government. He eventually got a CBE from the Commonwealth for his contribution to Commonwealth dducation.

He was a pianist and he organised a group. He had an engagement for his little band in Eltham on Saturday nights and he dragged me in to play with him. He was in teachers' college before me, with my brother. So I started to play in Dick Gardiner's band and besides our main engagements at Eltham on Saturday nights, we also played around Sunbury and the outer suburbs of Melbourne, Essendon, Sunshine and other places.

That had an interesting offshoot in that one night I met another ex-student of Melbourne Teachers' College, whom I knew was playing football with Eltham Football Club.

On this particular night, whilst I was playing on stage, he's merry and he came up in a drunken sort of way and introduced me to the captain of the Eltham Football Club and said that they needed players and praised me as a good football player. He persuaded the captain of the football club to sign me on.

So I played football with Eltham for two years which was quite high class football, because in those days they were in the VFL [Victorian Football League] thirds.

There were league firsts, league seconds and then there was this other group into which Richmond and Collingwood and all the league teams put a team. Eltham was the only team that was not harnessed to a league team.

It was very rough football because you had the young aspiring people climbing the ladder, to try to get into the league firsts. There were also broken down league players who were worn out on their way down and had slowed down.

The only way they could perform was with all the dirty tricks they'd learnt in the league. It really was the dirtiest football that I had ever played. After two years of it, I decided I just couldn't risk that sort of injury.

Talk about injury, when I was at university, I won the university boxing championship in 1932 in the lightweight division. When I came back to Melbourne and I was doing physics part two and part three, I started boxing again. I won the university championship and in 1936 won the inter-varsity championship also.

Then I became secretary of the boxing club and later president, and also for a while, when they couldn't get a coach, I coached and managed. After I'd finished my own career and stopped competitive boxing, I coached and managed the university boxing team and took them away on inter-varsity matches. This was in the 1937-40 period.

Was this while you were still teaching?

Yes. In order to practice, I used to go in and fight and train in a professional gym in Bourke Street. That was an horrendous business. Again I don't quite know why I did it. What was in me that made me do it I do not know? I knew I wanted it as a training pitch.

A wonderful experience it was too, because one bloke I remember I used to have to spar with was 'Spider' Webb. He was the State Lightweight Professional Champion. They put me in the ring with him as a sparring partner. I spent two rounds just trying to keep out of trouble and if he hit me hard, I would just lie down on the mat and call for the next training partner.

I used to be scared stiff every Tuesday and Thursday night and I'd go in there in fear and trembling. Why I went, I still don't know. Incredible the things you do when you are young and stupid.

Finally though, I gave up boxing because of the fear of brain damage. I fought on one occasion in one of the major fixtures in the boxing year at that time, which was the Railways Institute Tournament. It was the biggest tournament next to the Victorian Championships.

I entered to gain experience. I got as far as the semi-finals and I won a bout. But although I won, I took some terrible head punishment and had severe concussion. I was not able to front up for the finals after that fight and, actually, for six months afterwards I could feel the results of that damage. That frightened hell out of me, because as a scholar I could see if I went on any more, I could become 'punch drunk'. So that finished that.

WORLD WAR II, 1939-45

Getting back to the time you joined the RAAF, it was very short lived. You got back to Melbourne University. What were you involved in there for the next few years?

Well my professor had become the chairman of what was called the Optical Munitions Panel. When the war began, we found that our supplies of optical munitions from England had been cut off, particularly after Dunkirk. They couldn't afford or provide anything for us.

The Ministry of Munitions then set up this Optical Munitions Panel, to try and get an optical industry going in Australia. The aim was to produce some much needed stuff such as gunsights, rangefinders, dial sights and various other devices including binoculars.

It was a fascinating business, but I remember becoming very cynical when I found that the big optical firms in England wouldn't give us any of their 'know how'. Even in the middle of war and it was a matter of survival, they were afraid that if we set up an optical industry here, after the war they'd lose a market.

Professor Hartung, in the Chemistry School, was the key man in the development of producing optical glass in conjunction with a big company [Australian Consolidated Industries] in Sydney. It used to make glass bottles, but then switched over to producing optical glass.

The Munitions Supply Laboratories of Maribyrnong produced a lot of the specifications and did testing work. Meanwhile the Physics Department did all of the design of lenses and a lot of their testing.

The Botany Department set up the Graticule Manufacturing Laboratory, which made the eye pieces for optical instruments that had the cross lines or a graduated scale. It was very delicate work producing these.

The wartime optical industry really boomed. Within three years, we achieved in Australia, what in England, had taken over fifty years to develop. One of the tragedies in Australia, I believe, is that at the end of the war, they let the whole thing collapse and they didn't push on with the optical lens industry.

So for the years of the war, I was a research physicist doing all sorts of jobs in relation to optical munitions. Due to my administrative ability, I was made First Assistant Secretary and later Acting Secretary of the Optical Munitions Panel, which taught me a lot about administration.

The Optical Munitions Panel was a big high level committee with top quality people on it including public servants and professors. I learnt all about taking minutes, running meetings, organising papers and this massive administration as well as developing a knowledge of filing techniques, office procedures and things like that. That stood me in very good stead later when I became head of the Antarctic Division.

Towards the end of the war, when I had failed to get into the RAAF because of Manpower Regulations, I still had this yearn to get into the war itself. I had that as a sort of goal and I was determined to get there somehow.

What was the motivation?

Oh, just adventure. Finally I was able to persuade the Army to send me to New Guinea to conduct a scientific reconnaissance. The binoculars and gunsights in the tropical areas were degenerating extremely rapidly because of infection with fungus. The eyepiece would just get completely blocked up, the lenses would grow over with a network of fungal filaments.

We were doing research on this at Melbourne University and I persuaded the Army that I should go to New Guinea and do two things. First, a survey of the effect of fungal infection on all the instruments in the battle areas and secondly, carry out a number of tests on some of the devices we had thought up, and trial them in the jungle.

So early in 1944 I went to New Guinea and toured the battle areas. I wrote a report for the Australian Army on what was wrong with their workshops, how they were failing to cope with the cleaning and supply of the instruments and also carried out those experiments. I was thus able to get the war part of it out of my system.

You moved right through the whole combat area?

I went to most of the combat areas and saw much of New Guinea, more than most soldiers, I think. I didn't participate in any battles, but a couple of places I was staying at got bombed at night.

THE IMMEDIATE POST-WAR PERIOD

When the war finished, I reverted to being a lecturer at Melbourne University. An amusing point arose at that stage, at the end of the war and I had been away from the Education Department for six years. I had a Master's degree with Honours in Physics. I was a lecturer at Melbourne University and I was still on the Primary Roll of the Education Department. So I got a letter from the Education Department suggesting that perhaps the time had come for me to resign. So I resigned.

There was no PhD at Melbourne University when I went through, so a Master's was the equivalent. They had a very high level Master's degree, which took two years to obtain and it was about equivalent to what you do for a PhD these days. When they introduced the PhD, I'd already reverted to doing physics research.

I was doing research in classical heat experiments with a famous physicist called GWC Kaye, who had been working with Professor Laby on heat earlier. My professor at that stage was Professor Martin.

After the war, Professor Martin suggested to me, when the PhD was finally introduced at Melbourne, that I should enrol, which I did. I switched over then to doing cosmic ray research.

I completed enough work to present for my PhD but again time blocked me. The PhD was normally two or three years. If however you were doing it part time, as I was, as a university lecturer, you had to make it in three or four years.

Now at that point, I left the university to take on the Antarctic job. But I had not complied with the regulations to qualify for the PhD. I had completed easily enough practical work and I had written papers and things, but the university suggested that I should keep myself enrolled. It was hoped I could spare time from the Antarctic Division to come back later and spend one more year to comply.

So I stayed enrolled for the PhD for quite a number of years until I finally told them there was no earthly way I could ever come back for a year and finish it. So I didn't ever actually comply.

Eventually though, they gave me an Honorary Doctorate of Applied Science.

This brings us up to I guess, 1947 when you joined the Antarctic Division. During the war and post-war years, when you were at Melbourne University, was most of your work in research? What sort of workload did you have and what students?

I was a lecturer in physics and a tutor in Newman College in physics. I was a demonstrator in various practical classes in the physics department and putting in as much time as I could get outside of that, carrying out research of some sort.

I began with measuring normal conductivity of gases and then I switched on to cosmic rays. It was when I was doing the cosmic ray work, that Melbourne University was asked to support the newly-formed Antarctic Expedition, by establishing a program in cosmic rays for installation on Heard and Macquarie islands.

The cosmic ray program was under the general leadership of a Dr Rathgeber, a German. But when this expeditionary talk arose, I was put in charge of the production of the equipment which would be sent to Heard and Macquarie islands. It is interesting to note three of the people involved.

David Caro, the chief designer of the electronic gear, who later became the Vice-Chancellor of Melbourne University.

Fred Jacka, one of the men working in the team, later became my Chief Scientist and then the Director of the Mawson Institute for Antarctic Research in Adelaide.

Thirdly, Ken Hines, one of the men who went down with the cosmic ray gear to Macquarie Island, later became a reader in physics at Melbourne University.

It's interesting how it all developed.

CHAPTER TWO

HIS ANTARCTIC SERVICE

So you were approached personally?

Well I had better say how I got into the Antarctic expedition business. Due to my interest in snow, ice and skiing and the fact that I had read all the books about polar exploration, I had developed a rough ambition one day to go to Antarctica.

Then one day, I had heard rumours that the Australian Government was setting up an Antarctic expedition. There was nothing in the newspapers, I couldn't find anything about it, there were no advertisements. I was on the point of writing to Sir Douglas Mawson, whom I'd heard about but never met. I thought that if someone was thinking up something about Antarctica, then he would be the bloke who'd probably know about it.

I was nearly about to write to him and ask what was going on, when I was walking down the passage of the Physics Department with Professor Leslie Martin. He was the Chief Advisor on Defence Science to the Australian Government and he used to go to Canberra once every two or three weeks for meetings.

Just out of the blue he said, as he was walking down the passage with me, 'Oh Law I've just come back from Canberra, and we're having trouble finding a chief scientist for this Antarctic expedition.' I just couldn't believe my ears so I said, 'Did you mention my name?' He said,

'Surely, Law, you wouldn't be interested in that?' I said, 'I'd give my right arm to go on that expedition!' He said, 'Goodness me, I'll go and ring up.'

So he went and rang up and within a few days, I was put down for an interview. Then, within a few weeks, I was appointed Chief Scientist for this new Antarctic expedition.

It was 'The Australian National Antarctic Research Expeditions (ANARE)'. The head of it was a man seconded from the Department of Civil Aviation, who was an RAAF Group Captain. He had been the pilot in Mawson's BANZARE expedition in 1929-31. His name was Stuart Campbell.

I was made his number two, to look after the scientific side of it. I was then also put in charge of cosmic ray work. (Not in charge of the cosmic ray work altogether because Rathgeber, was the director of that. But in charge of the production of the equipment that was to go to the island stations.)

So we had a small team, there were: Hines and Speedy, the two who were going to Macquarie Island; Jacka and Jelbart, who were going to Heard Island; and there were McCarthy and myself, who were all going to sail in the *Wyatt Earp*.

One thing I had organised was to build a hut at Mount Hotham, as a testing place for CR equipment. Then to coordinate our efforts to get all our CR gear up to Mount Hotham and into the hut, test it and get it back again, fighting against time to it get ready by the end of the year. Putting it all onto the ships was quite an episode in itself, full of adventure, excitement, fun and hard work.

All those people you worked with were basically from Melbourne University staff?

Yes.

Could you perhaps tell us a little anecdote about the pre training at Mount Hotham?

One amusing anecdote is that I wanted to get this hut built at Mount Hotham. I had to apply for occupancy permission from the State Department that ran the Hotham area; but I could see that if we didn't get the hut built before April we'd never get it built because of the snow.

So I went up to Bright and met a builder and we designed a little hut the way we wanted it. I got the builder to get to work on it and finally the Department concerned said they wanted to send someone up to look at this site. So I took the bloke up and when we got up there the hut was already finished.

The Department was angry, but I pointed out that the stuff was to go away the following December and if the hut hadn't been built in April, there was no earthly way that it'd be finished.

As it was, I remember, just towards the end of the hut-building episode, we were up there with a Government car and it got snowed in and we all had to walk out. The Government car driver with a long overcoat on, nearly collapsed and we had to drag him, because he was not fit enough to be tramping through six miles of snow. They had to dig out the Commonwealth car six months later at the end of winter.

You relayed to me quite some time ago a story about nearly dying of hypothermia in that region which was the closest you had ever come to death.

Oh yes, that was in about September, when I took Bob Dovers, one of the Heard Island party, up to give him experience of snow, which he didn't know anything about, and also to do some jobs at the hut. On the way out from Hotham, we got caught in heavy rain, got soaked and then had to ski up over the top of Mount Hotham to get out.

At that altitude, the rain had turned into snow and the temperatures were sub-freezing. All our sodden clothing just iced up and no matter how hard we worked going up hill, we found we were just getting colder and colder. Going down from the summit we called in at the Diamantina Hut where some skiers had a fire.

Dovers was blue and shivering uncontrollably, so we stripped off his clothes and massaged him in front of the fire. An hour later we resumed our trek down. I was not so bad (warmer clothing). Only for the hut we would have been in desperate trouble.

Phil can you perhaps give a little background into what your role was in that first year after you had joined ANARE?

There was no Antarctic Division as such at that time. Herbert Evatt, who was the Minister for Foreign Affairs, had set this expedition going. He couldn't sell the idea to any of the Government departments, not even CSIRO wanted to pick us up, so he rather reluctantly had to put us in his own Department.

So we were a little section loosely tagged on to the side of the Department of External Affairs and we were directed by what was called the Executive Planning Committee.

It was a very high-ranking committee consisting of the heads of Departments that were involved, including the Bureau of Meteorology, the CSIRO, the Bureau of Mineral Resources and sections of the armed services.

The Committee also included Stuart Campbell, the man appointed as leader of ANARE, as well as Sir Douglas Mawson and Captain JK Davis. When I joined the group I became a member of the Planning Committee too.

I joined as Chief Scientist and my job was to prepare a rather complicated scientific program for the first year's operations. That involved liaison with the universities, the Bureau of Meteorology, the Ionospheric Prediction Service, the National Mapping Office and the Bureau of Mineral Resources.

In addition, I was involved in the logistics of getting ready for the three particular voyages: one voyage to Macquarie Island; one to Heard Island; and one in the *Wyatt Earp* to the Antarctic continent. To provide all that was required, for the two island stations plus the voyage of the *Wyatt Earp*, meant a tremendous amount of work, because the headquarters organisation consisted only of Stuart Campbell, an executive secretary Trevor Heath, a storeman George Smith, a couple of typists and myself.

The only other help we had consisted of the expeditioners as they came on strength. They were sent out to our store at Tottenham to work on getting everything ready.

So it was during that particular period in September that I took Bob Dovers up the mountain and had that experience that I just recounted.

So your first set of expeditioners joined you in around about August of 1947?

Yes. As well as this, although I was seconded from the physics department in Melbourne University, I had not left there. I was an employee. I was given a year off without salary, but I was still project leader for the three bits of cosmic ray equipment which were to go, two to the stations and one on the *Wyatt Earp*. I was to look after the bit that went on the *Wyatt Earp* with the assistance of the man from the Bureau of Mineral Resources, Ted McCarthy, who was a geophysicist.

Another job I had was to design and have built a laboratory which was to be perched on the boat deck of the *Wyatt Earp*. As the ship was over in Adelaide this involved a certain amount of liaison with the Royal Australian Navy.

That first year must have been pretty hectic for everyone, preparing for something that was brand new.

Yes.

Selecting the personnel for that first adventure must have been difficult. How was that achieved?

It was done by a system that I completely disagreed with. They made no attempt to advertise the jobs or to cast the net wide throughout the whole of Australia, which I felt was terribly important for a national event like that.

What they had done was to use the 'old boy network' you know, men on the Planning Committee would say

'Hey Joe do you know someone who might be interested?' And Joe would say, 'Oh yes I know of a nephew, he's very good and he's been in the Boy Scouts', or 'my friend Tim has got a son who's a bushwalker and he's very good.'

So they went about it like that and Stuart Campbell was on the selection committee and he was always very impressed with what we call tough, aggressive characters. He'd regard them as having strong character with capital 'S' and 'C'.

He was completely unaware that if he put fourteen or fifteen of those sorts with an aggressive macho character together in some secluded spot, it wouldn't work well.

Also they hadn't bothered to really check referees and we got some rather peculiar 'no hopers' amongst the lot. However by and large, though, it didn't work out too badly. The ships were the problem. There was the *Wyatt Earp* which... Did I say anything about that earlier?

No you haven't spoken about that yet.

Well when they were searching for a ship, they approached the Navy. The Navy, very reluctantly agreed, after a lot of discussion, to loan one of the LSTs they had, LST 3501, which was a relic of World War II. They agreed to make it available, but they were afraid it might break in half on the voyage, so there was a bit of apprehension about it.

In any case, it set off and did the job. It went to Heard and then Macquarie islands. Incidentally it didn't break in half, or nearly break in half until about four years later, when I was in it and everything broke down.

Yes, after the first year when I got back safely, they

seemed to forget about worrying about it and of course, that was the very time that they should have started to worry.

Those landing ships tanks (LSTs), they weren't particularly well suited. Can you tell us about them?

They were about three and a half thousand tons and they were very long, about three hundred feet nearly, I suppose. The superstructure was all aft with the engines. The long tank deck ran right out in front, like a slapping springboard, and in the bows there was a door that could be lowered at the beach.

The well deck held the tanks and on each side of this tank space, as they called it, was the accommodation for the crew and the men for the expeditions.

It was pretty horrible in there because it was not insulated and the bulkheads or walls used to drip with moisture and condensation when it got cold. There was no insulation for sound and you could hear every wave bashing the side of the ship.

The ship itself used to make a lot of din because when you went over a wave and hit the next one it was like a thousand kerosene tins crashing. The whole front of the springboard would just flap and go 'yong yong yong yong' and the sides of the bulkheads would just flap in and out and make 'kjonk, kjonk, kjonk' noises.

If you were standing on the deck as this wave and ripple went past it would throw you off your feet. We used to play deck tennis down on the tank deck and it was almost impossible to keep your feet as these ripples went through underneath.

You can imagine with all that flexing, it was no wonder that ultimately it cracked up.

The *Wyatt Earp* was a different story. It had been a sardine ship in Norway and was bought by Lincoln Ellsworth, who was an American Antarctic millionaire explorer. He renamed it and took it down to Antarctica on two or three voyages with the Australian, later Sir, Hubert Wilkins as his advisor.

After the last time, which must have been just before World War II, he was so fed up with travel in this ship, which rolled abominably, he said that he didn't want to see it again and he handed it over to Wilkins when they got back to Australia.

Wilkins, smartly, sold it to the Australian Government for ten thousand pounds. Then war broke out and the Australian Government used it to run explosives between Brisbane and New Guinea.

At the end of the war, they didn't know what to do with it, so they moored it in the Torrens River at Port Adelaide. It was used as the headquarters for the Sea Scout organisation.

It was then, when the expedition was being set up, that Mawson in Adelaide, supported by JK Davis, mentioned it and persuaded the Planning Committee to take the *Wyatt Earp* on the first reconnaissance voyage.

This voyage was to go down near Mawson's old base at Commonwealth Bay and have a look at whether there was a site that could be used for a permanent station somewhere around there, or at Cape Freshfield, which was further to the east.

A plane was taken on board the ship to do some aerial photography of the area as well. We were also to visit the

Balleny Islands, which had been visited once in the middle of last century and were imperfectly mapped. So we were to do a survey of those.

We were to take geomagnetic measurements whenever we landed anywhere. So we wrapped up a program. There was meteorological equipment on board, a bit of plankton catching, some sea water temperature measuring bottles which we lowered and so on.

Did it have any lab facilities at all that were utilised in these early days?

Yes, this lab which I had built on the upper deck for the cosmic ray equipment. There was nothing else. The meteorologist simply worked out of his cabin.

Can you describe the Wyatt Earp *for us?*

The ship was made completely of wood and was about a hundred and thirty-five feet long, very small. It was smaller than Captain Cook's ship. It had no bulkheads, but it had very thick timbers in the hull and bows. Nearly three feet thick, it was amazing.

It had no bulkheads because they reckoned if it filled with water it would probably still float. It might have in the early days, but not with the big diesel engines that we put into it. These large engines later turned out to be our undoing, because they produced a sag in the boat, which produced terrific tension on the tail shaft.

As you know, in a ship the tail shaft is connected directly from the engine to the propeller. There is no universal

Senior Scientific Officer with ANARE at the Albert Park Barracks, 1947
Phillip Law collection

Phillip (left), man-hauling a sledge over sea ice, Prydz Bay, 1955
Phillip Law collection

Nel Law christening the *Nella Dan*,
Copenhagen, 1962
Phillip Law collection

Phil and Nel, Melbourne, 1940s
Wendy Law Suart collection

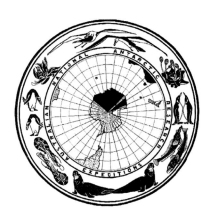

ANARE logo designed by Nel Law,
1950
Australian Antarctic Division

ANARE logo featured on a stamp
issued for Antarctic Research in
Antarctica, 17 November 1954
Australia Post

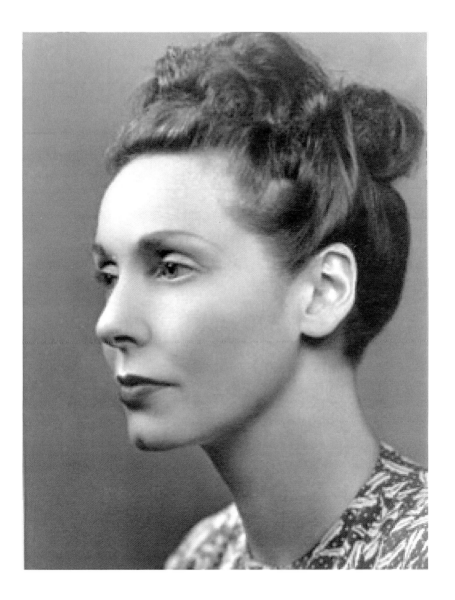

Nellie Isabel 'Nel' Law (nee Allan), 1950s
Phillip Law collection

Phillip, with Steve Crabb (left), Victorian Minister for Conservation and
Environment, viewing a demonstration at the
opening of the Queenscliff Marine Station, February 1991
Phillip Law collection

Phillip speaking at the Victoria Institute of Colleges' first degree
conferring ceremony at the Pharmacy College, 1969
Phillip Law collection

joint and the result was, as the engines' supports sagged, the tail shaft bent till it was on the point of snapping.

So did the whole hull flex under the section where the engines were?

We couldn't understand why the engines had sagged. Admittedly they were very heavy and the engineers on board said, 'Oh well they have sunken in their wooden supports.'

Now as a physicist, I knew that was impossible. You put an engine on wooden supports and it is supported over such a wide area that the pressure per square inch is not great enough to even dint the wood, let alone sink into it.

So we had to look for another explanation, which wasn't forthcoming until we returned the following year. Then we found that in its early history, the *Wyatt Earp* had been imprisoned for a year in the ice somewhere up in the Arctic. It had been under pressure and had been lifted up with ice beneath the keel. It sat there for four to six months through the winter being supported on its keel and this produced a bow in the keel.

Now that was all right when it was just a sardine boat. However when they mounted those heavy diesel engines in Australia, that weight acting over a period of six months, just gradually flattened out the keel bend to where it was originally. It was only a matter of a few tenths of an inch, but it was enough to threaten to break the shaft.

I remember we got about three-quarters of the way to Antarctica on the first voyage and this was being reported to Navy HQ. The Naval Board, as it was called then, finally said 'You've just got to come home. That thing will snap very soon.'

I remember all of us were furious. There wasn't a man in the ship who wouldn't have gone on and taken the risk. We talked about the shiny bums back in headquarters who didn't know what the real situation was and so on.

But how right they were, I mean in another week or two the tail shaft would have snapped and there wasn't another ship nearer than America. No one could have possibly come to rescue us. Anyway, it was a very hairy situation altogether. I could talk all day about the discomfort and the other points about the *Wyatt Earp.*

Perhaps a bit about the accommodation and an average day on board?

The accommodation was built down each side of the superstructure and between that and the central section there was a passage on each side. The central section was almost completely devoted to the engine-room spaces, but adjacent to the engine-room casing there was a little cabin for Stuart Campbell, Ted McCarthy and me. It had no scuttles or portholes because it opened out each side onto the passage and it only had a door onto one side of the passage.

There was a revolving twist-plate ventilator in the boat deck above. Of course that meant that water came onto the boat deck when it rained or snowed and this just dripped through onto my bunk. So we had to plug that up. That meant that we either kept the door shut and produced a fog that gave us a headache overnight, or we opened the door for ventilation and got all the diesel fumes from the engine-room coming in. So we always opted to keep the door shut.

Stuart Campbell had a thing about living tough, never washing and not washing his clothing or pyjamas or underpants. Towards the end of a few weeks it was pretty high in our foggy cabin at night.

Our cabin was one of the few that was dry on the ship, because when they re-built it in Adelaide, they had to add several accommodation cabins. In doing that they blocked up a number of scuppers on the previously open deck and built the accommodation over it. They didn't block up the scuppers properly though and when the ship rolled enough to dip the gunwales, the water came through.

This wasn't detectable when it was in dock because it wasn't rolling, but the result was that every cabin on that level, which was everyones' except the captain's, the chief engineer's and first officer's, had three to four inches of icy cold water slopping around on the deck.

Our little cabin was one that didn't leak either, because it wasn't abutting any outside surface. But that didn't last long because the fellows immediately astern of us, whose cabin opened directly onto the stern of the main deck, got a brace and bit and drilled a few holes through the bulkhead to our cabin. As their cabin was a bit higher than ours, all their water drained through into our cabin.

Poor Dr Loewe, the meteorologist, used to take weather observations every six hours and he'd get up in the middle of the night to take his met obs. He'd jump down, in his bare feet, out of his bunk into three or four inches of slopping freezing water.

We lived like this until we got into the pack ice. On one occasion we had a reasonably calm day and the first officer,

Bill Cook, took some crew over the side on a rigged-up platform, to nail a canvas patch over part of the side of the ship to try to keep the water out.

The ship was rolling enough to be dunking them almost over their waists on every roll. They'd last for about five minutes nailing the canvas and they'd have to switch over to someone else, as it was freezing.

We'd been short of water and that was the worst hardship, as there was no way of producing water on board and we could carry very little. We longed for the time when we got to the pack ice, so we could melt ice and get some fresh water. Water was rationed. Apart from drinking and cooking, we had one pint of water each for a whole day. You used that to wash the tips of your fingers and the sleep out of your eyes in the morning and clean your teeth. That was all.

When we came back from down south after the tail shaft trouble erupted, we didn't dare go to Hobart where we were given such a rousing send off. We were too ashamed to crawl back in there. So we crawled back to Melbourne, where we spent two or three weeks being overhauled and having shims put under the engines to level them up.

Finally in February we departed and went down again. Of course that was so late in the summer that by the time we got down there, the ice was freezing. We got to within about thirty or forty miles of the Mertz Glacier Tongue, near Mawson's old station, and had to turn around and come back.

The plane made one flight. That was a bit of a shambles too because no one had been properly drilled on how to sling it out and back on deck again. This was a very hazardous operation.

It must have been rather frustrating to get so close and have to turn around again.

Yes.

How far did you actually get on the first voyage?

About three quarters of the way down. We were a fair bit further south than the latitude of Macquarie Island. On the second trip we did one valuable thing, a running survey of the Balleny Islands and actually made the second recorded landing on them.

The man who discovered the Balleny Islands was Captain Freeman in the ship *Sabrina* back about 1840. He was sent out by the Enderby Brothers Whaling and Sealing Company just to explore. He made a short landing.

It was a pretty impregnable place and very difficult to find any sort of beach to land on. The cliffs are up to a thousand feet high, with three or four hundred feet of ice on top of them. The islands are glaciated, very much like Heard Island, only worse.

Freeman got ashore for a couple of minutes and had to beat a hasty retreat, so Stuart Campbell, the leader of our expedition, was very keen to get ashore.

We took a whaleboat and rowed in and searched for a possible landing place. We found a small beach about twenty yards wide, where there were big boulders and volcanic rocks, shingle, no sand and great blocks of ice from glaciers that had been washed up. Altogether it looked pretty forbidding.

The wind was blowing, it was cold and the water was about thirty degrees Fahrenheit. We rowed in and three of us nearest the bow jumped out. Stuart Campbell went first, I went second and then a seaman jumped in, I've forgotten his name. We tried to hold the boat in place while Stuart Campbell rushed up the beach and grabbed a rock specimen. But the waves were bashing over the stern and threatening to broach it and push it side on, so everyone started to scream at us to come back.

We pushed off, climbed in and in a snowstorm, rowed our way out two miles to the ship, which had backed off hurriedly. We got back on board, absolutely frozen and were very excited about having been the second party in history to land on this island.

Did you get your rock?

'Stewy' got the rock and brought it back, yes.

We had a terrible storm on one occasion at the Ballenys, a real hurricane, my first experience of something over a hundred miles an hour. Luckily we were able to get in under some steep cliffs on one island and shelter.

We steamed up and down about three quarters of a mile in the lee of this small island and luckily we did, because I think we'd have had a pretty rough time surviving in that hurricane.

After that we had one perfect day which enabled us to sail around the whole complex of islands, and chart them by radar and produce a very decent sort of map of the whole lot.

How did the Wyatt Earp perform in really heavy seas?

Well if you turned her bow into the wind she rode pretty

well, but she rolled abominably. She had the worst roll of anything I'd ever been in or heard about. She'd roll forty to fifty degrees each side of the vertical with a total swing, that is one side across and back again, of four and a half seconds.

Now that produces an angular acceleration that is so violent that nothing stays put. Cupboards actually burst the locks that are locking them, stuff gets hurled away. Nothing ever falls off a table, if the rolling is like that, it just travels through the air six to eight feet to one side and smashes against the bulkhead.

We had a bad storm in Bass Strait just after we left Melbourne on the first trip. If we hadn't been able to anchor in the lee of Flinders Island, I think we'd have been wrecked, because the whole ship literally went out of commission. Everything broke loose. No one was able to carry out normal duties.

The radio room was an absolute shambles, everything had come adrift. The radio sets, the batteries, the typewriter and all the typing paper, carbons, all the stationery and all the forms, all sloshing round in six inches of water on the deck. The radio operator had a packet of powdered soap for his laundry which had somehow burst, so everything was frothing with this powdered soap.

Well you got back to Australia that summer after those two trips and didn't achieve what you had really set out to do. What happened then?

I went back to the university to write up my cosmic ray results. I was keen to continue those cosmic ray observations, which were measuring a latitude effect and continue those over the equator. So I found out that the Army was sending a ship to Japan. The *Duntroon* was a ten thousand ton passenger ship and was well known on the Australian Coast.

It was to pick up Australian troops and bring them home. That would have been August 1948. So I persuaded the Army to let me go on board with my cosmic ray gear. Hence I did a return voyage to Japan, while I was still with the Antarctic Division.

When I'd got back from that, I'd spent my twelve month leave, so I went back to the university again as a lecturer.

Between August and September that year, Stuart Campbell decided that he'd go back to his permanent job, which was Director of Air Navigation and Safety at the Department of Civil Aviation.

His reason for doing it I believe was twofold. First, he was having a lot of trouble with the rather stodgy bureaucrats and administrators in the Department of External Affairs and, secondly he had tried all over the world to find a ship to replace the *Wyatt Earp* without success. He couldn't see much future in just going up and down to Heard and Macquarie islands.

So at the end of that year, he returned to his post and I was appointed to take his place as leader. Being a physicist, I could see a big future in Heard and Macquarie islands. Each of them was a very fine place for geophysics and science generally and while I was waiting for a chance to get to Antarctica, I was quite prepared to spend several years building these stations up. So that is what I did from January 1949.

Incidentally, at the end of 1948 they also formalised the situation of this funny little group tagged on to External

Affairs. They made it a Division of the Department of External Affairs and I became the first Director of the Antarctic Division. That occurred in January 1949.

At that stage, once the Government's attitude to Antarctic research had been formalised by setting up the Division, what sort of facilities and staff did you have at your disposal?

We still had the small staff that I mentioned. There was my offsider, Trevor Heath and myself. There was a woman who looked after the personal cables back and forth from the men, there were about three typists, and a storeman out at Tottenham. That was all.

Well, I began then, drawing up lists of the sorts of men that I wanted to employ in the Division and I redesigned it all. We had been quartered in an old bluestone building at Victoria Barracks in St Kilda Road. That was quite ridiculous, it was only really two rooms.

So I was able to negotiate with the Navy to give us a block down at Albert Park Barracks in Block H, which was staffed by a lot of Navy clerks. I got the Department of Works to do this out in some nice modern light colours, instead of the old stale brown they all had, and we made quite an attractive place of it.

Then over the next few years we built the Antarctic Division up gradually with the appointment of permanent officers.

The first big job we had was to take charge of our own provisioning. Until then, all equipment, except food, had to be purchased through the purchasing system of the RAAF. It was quite hopeless because sometimes you'd lodge an order and it would take a couple of months before you could get it.

Firstly, they'd notify every store in Australia and ask if they had it and that could take six weeks. If they didn't have it, you were allowed to put an order out with a private firm. With orders coming in from Heard Island, sometimes only a week or two before a ship sailed, this was quite impossible.

The other thing that was unsatisfactory was the food situation. The suppliers were provided by the Navy and they were provided according to Navy victualling. This was quite unsatisfactory for our purposes, so I set up a supply section to do our own purchasing and accounting. We had to design all our own purchasing forms and receipt dockets and everything else, sort of starting off from scratch.

I am glad that we did it, as it made a tremendous difference to the efficiency of our organisation. The way we could hustle stuff through in purchasing was quite remarkable for a public service organisation.

Our blokes would make good contacts with the Contracts Boards' people in the Department of Supply, the RAAF, the RAN, William Adams Tractors and various other people.

Often a telephone call would be enough and we'd say look, the paperwork will come along later, but we want so and so, can we come out and collect it tomorrow? This worked wonderfully well.

You obviously established good credibility which stood you in good stead for years to come. The first year that you had a wintering party at Heard and Macquarie, when was that?

In 1948. That was while I was in Japan.

So that group of people that you went down with on the LST stayed there?

Yes. Now I hadn't been to Heard and Macquarie islands because I was getting ready for the *Wyatt Earp* voyage. Stuart Campbell went to Heard Island to set it up and the commander of the LST, Lieutenant Commander George Dixon, went down to Macquarie to set up the station there.

Stuart Campbell and I came back to Macquarie Island in the *Wyatt Earp* on the way home. So by the time we got back I had seen Macquarie and Antarctica and Stuart Campbell had seen Macquarie, Heard and Antarctica. My first trip to Heard was in January 1949 in the LST.

From 1949 you were involved in searching for a ship that would enable you to set up a base on mainland Antarctica.

Yes, two things of consequence happened as well as running these two stations. In 1950, I persuaded a new European Antarctic expedition to take me as an observer, so I could get some Antarctic experience.

This was the Norwegian-British-Swedish Antarctic Expedition (NBSAE) of 1950-52. It left from Cape Town in a little ship called the *Norsel*. This was a kind of patrol boat left over from the war, that some Norwegian entrepreneurs had refurbished. They were letting it out for sealers and others in the Arctic and this NBS expedition had chartered it.

Was it smaller than the Wyatt Earp?

It was bigger than the *Wyatt Earp* and much better. It had bulkheads, powerful engines and had a sloping bow, which had been ice strengthened, and could break through reasonable ice. It wasn't a bad little ship and I had to go to Cape Town to meet it.

I went down with them just on the summer voyage, so I got back to Cape Town at the end of March, early April. I had arranged three things which were important. First I planned to go on from there to England, to see the Antarctic experts in London and Cambridge to buy special polar gear for Antarctic work for the stations.

I persuaded the NBS expedition to sell me their two Auster aircraft when they had finished with them, which was to be twelve months later. I also arranged to buy one of the huts, that they had built in Norway. They had to offload it at Cape Town, because they didn't have room in the *Norsel* to carry it.

So by the time I got back from London in mid-1950, I'd had some Antarctic experience and picked the brains of all the British Antarctic blokes. I had also been around all the suppliers. I'd gone to Norway as well and thus had contacts for buying anything I needed from Norway or Britain. I had the two Auster aircraft and one supply hut on tap as soon as I could ship them back to Australia.

While I was on that trip I was looking for a suitable ship. I went to the USA also, but like Stuart Campbell, I couldn't find anything around the world. When I got back to Australia in late 1950, I worked with the chief naval designer on the Australian Ship Building Board to design an Antarctic ship. It took us two years to finish the designs and they were completely finished, right down to the furniture, hangings, crockery and everything else.

It would have been a fine little ship, along the style of the *Kista Dan*, although it would have been more commodious than the *Kista Dan*.

Just when I was at the point of putting the proposal up to the Government, I read that the Danish firm of Lauritzens had built the *Kista Dan* for trade on the East Greenland coast in the northern summer.

It struck me that they wouldn't have anything to do with it in the northern winter and probably just have it tied up. So I wrote to Lauritzens and asked if they would be willing to charter it to us for the northern winter, our summer, and the time we needed it.

They agreed and that was the beginning of the long *Dan* ship history, that went on for almost fifty years.

One of the reasons for opting out of building our own ship here was that it was much easier to get funds, so many thousands a year, out of the Government to charter a ship, than it was to get several million in one lump to build one. Secondly though, I was afraid of the difficulties we might have in crewing a ship.

The Navy, after its experience of the *Labuan* breaking down, had decided it would opt out of the whole business, they were very reluctant to come in again and man an Antarctic ship.

I knew that with the ordinary mercantile marine we'd have every possible difficulty with the sailors and the unions, quite apart from the fact that we didn't breed good open sea-going people in Australia. They just go up and down the coast in relatively calm waters. They don't really learn what seafaring's all about, in the sense that the Norwegians and Danes do.

A couple of other things are worth mentioning about that period when the *Labuan* broke down. At the beginning of 1952 we were left stranded. We'd finished the Heard Island trip, but we still had Macquarie Island to do.

I had to arrange with the Australian Shipping Board to get a 'river ship.' These were built as ore carriers to carry iron ore from South Australia to New South Wales for BHP. A 'river ship' was a ten-thousand tonner.

Our's was called the *River Fitzroy*. It had no accommodation for passengers and we had to tear some stuff out to put in bunks.

It was a horrible journey to Macquarie Island for everyone. The captain and crew were scared stiff and they got one side of the ship all bashed in on the way back with a big wave. It was quite unsuited to the sort of job. But it enabled us to relieve the Macquarie Island Station.

Then we sub-chartered, from the French, a little ship called the *Tottan*. The French had set up an Antarctic station in Adelie Land and they were using the *Tottan*, which was smaller than the *Norsel* and not nearly as good. That is another story in itself, going on that little ship. The French would take their men down in it and then they'd give it to us and we'd go to Heard and Macquarie. Then we'd give it back to the French, who'd take it down and then pick their men up again.

The station in Adelie Land was to be permanent, but it burnt down, so they pulled out and didn't go back until the International Geophysical Year (IGY). When they did go back, they went to a different place.

So our experience with ships was very varied and the

54

step up from the *Wyatt Earp*, to the *Tottan* into the *Kista Dan* was the greatest luxury. It was quite a remarkable experience for those who'd been in those earlier ships. Changing from the *Kista Dan* to the *Nella Dan* was another step up. Luckily I never had to go backwards by stepping down in the quality of shipping.

That virtually brings us up to 1954.

Yes. The important thing was getting approval from the Government to actually set up Mawson Station. The first nail in the deck, you might say was when I found I had the ability to charter the *Kista Dan*. I was then able to go to the Government with a plan and say, 'Look, I can get a ship, this is what I want to do. Can you give us approval and the money?'

The essence of this was that it had to be very cheap. I had a plan whereby now the French had opened up a station on Kerguelen Island, which is just to the north of Heard Island. Whereas Heard Island was a vitally important meteorological station, we reckoned that once Kerguelen had started up it was only a hundred and eighty miles north and it could serve Australia instead of Heard Island. We could then pull out of Heard Island.

So I said 'We'll pull out of Heard Island if you let us set up in Antarctica. If we do that, we can transfer the huts from Heard Island with the diesel engines, plus the radio sets. The new station need only take nine or ten men. It would be smaller than the fourteen men at Heard Island and it wouldn't cost too much.'

On that basis we persuaded the Government. So in 1954 we were able to get Mawson off the ground. Then in 1955 we pulled all the men out of Heard Island and brought them home.

So the closing of Heard Island wasn't because you'd finished all the research you'd wanted to do?

No. It was a trade off.

You mentioned that everything Heard had been used for could be done at Kerguelen. What did you mean by that?

Most of the meteorological data necessary for Australian weather forecasting could be reported reasonably well from Kerguelen.

What was being done at Heard Island?

The usual things, meteorology, geomagnetism and seismology. The local things were important, like geology and surveying, but we had already done a reasonable job on those, during the five years we had been there.

The IGY didn't happen until 1957. By then we had already been running geophysics programs at Macquarie Island and Mawson, so were fully equipped and well into the programs. We had tried and tested all our equipment and we were sitting pretty. No other nation had full geophysics programs before the IGY, except us.

What other nations actually had bases in Antarctica at that time?

Eleven of them went down for the IGY. When we set up at Mawson, the only other ones down there were the British, Argentineans and Chileans. They were on the Antarctic Peninsula, right across the other side of the continent.

Then it was in either 1955 or 1956 that the French opened up Dumont d'Urville Station. For the first year of the IGY the Australian station at Mawson operated better than any other Antarctic station because all the others ran into teething problems which we'd already solved.

We were doing better with our little team than the big Russian and American teams and all the others. That didn't last for long. Once they overcame the first two or three years of problems they left us for dead. It was nice to feel that in the early period, we were more than matching them.

To what extent were you or your Department or the Government involved in the initiative to look at Antarctica as part of the IGY?

The decision to include Antarctica was made by the big world IGY Committee, which was organised by the International Council of Scientific Unions (ICSU). It set up an IGY Committee.

Australia then set up an IGY National Committee. The setting up of that was a bit of a shambles here. It was given to a famous physicist in Sydney to organise, but he didn't really get around to it.

I could see time going and going and I finally had to take the initiative and write some letters and demand that something be done and kick someone into having the first meeting.

Then it was not only Antarctica, but was the whole IGY in Australia. They had Professor Webster, who was a physics professor from Queensland University. They released him half-time from the university to be a sort of executive secretary to the Australian National Committee and do all the legwork.

He found it so overpowering and time-consuming that he couldn't cope and I spent a great amount of time between 1955 and 1956 helping Webster get this whole IGY national program organised.

So that was an extra load over and above your normal administrative and command responsibilities. Did you get to Antarctica for the change overs in 1955 and 1956?

I went down in 1954 to Mawson Station. As you know I went down in 1955 and explored and found the site where we later were to build Davis Station. I went down in 1956 and explored the Wilkes coast. Then in 1957 I went down and established Davis Station.

So the whole thing went something like this. In late 1947, perhaps we'll call it 1948 to keep in line with the others, I travelled to Antarctica in the *Wyatt Earp*.

The beginning of 1950 saw with me the Norwegian-British-Swedish Antarctic Expedition, down in the Weddell Sea area.

Then I was in: 1954, at Mawson; 1955, exploring Prydz Bay, which is where Davis Station is now; 1956, exploring Wilkes Land; 1957, establishing Davis Station; 1958, Prydz

Bay again, with more exploration there and also Wilkes Land; 1959, Oates Land; 1960, Wilkes Land; 1961, Oates Land; 1962, Oates Land; and 1963, the Amery Iceshelf.

In 1964 I went to Macquarie Island and sent my deputy down to Antarctica. In 1965 I explored Enderby Land and in 1966 we had no time to carry out any exploration and we lost the aircraft we were going to do it with anyway. So in 1966 it was simply relieving Mawson, Davis, and Wilkes then coming home.

Phillip, getting back to ships. You've travelled in quite a few in the years you've been involved. Would you like to make a few comparisons?

Yes, I was lucky in the sense that I seemed always to be moving up the scale. So although I knew nothing about ships at the beginning and was quite content with what I had. Once I went up the scale and found how good a ship could be, I'd have hated to go back down the scale again.

You see we started in this little wooden ship the *Wyatt Earp*, which, really was a terrible ship to have to go to sea in. But I didn't know. I put up with it and thought that it was normal. It wasn't until I found what ships were really like that I realised what a dreadful little tub this thing had been.

It was smaller than Captain Cook's ship and far smaller than Nelson's ship. It was only a tiny thing made of wood. We lived on hard-tack out of barrels, bad water and bad meat, all the problems of the ancient sailors.

Talking about bad food, did you ever have a problem with scurvy?

No we knew enough about scurvy to have fruit juices and tinned fruits and things of that sort. There were no vitamin pills in those days.

The other ship we had at the same time, right at the beginning was the *Labuan*. In the beginning it was just called LST3501. It was one of those tank-landing ships with the blunt bows and the doors that opened up at the front to run the tanks out. It was quite unsuitable for the terrific waves and winds that you get down south.

As a matter of fact a lot of them broke up, during the War, crossing the Atlantic. The ones that broke up were nearly all welded ships. Luckily, ours was riveted. I don't think the Navy would have sent a welded ship. Even so, ours broke down in every possible fashion on its last voyage in 1951 or 1952.

The first thing that happened was that the plate rivets loosened and sea water got in. It then mixed with the fresh water and the fuel. The continual buffeting had shaken the condensers loose on their bolts enabling the sea water to enter them. The condensers were heated by steam and were then to cool the steam. But heating the salt water resulted in salt precipitation which then blocked the condenser pipes.

As a result, the steering gear failed and they had an engine-room fire. The engines had stopped, of course, and with them, the electric lights went out. We were three hundred miles from Fremantle completely derelict, with no lights, water, or power. After two months of incessant storms, we had three days of absolute calm, so they sent a tug and brought us in. Now if we'd had one more decent gale we'd have had it.

Talking about breaking in half, a great crack developed

HMAS *Wyatt Earp* on ANARE voyage,
8 February - 1 April 1948
Laurie LeGuy photo, Australian Antarctic Division

HMAS *Labuan* (LST 3501). First ANARE voyage to Heard Island,
during stopover at Port Jeanne d'Arc, Kerguelen Islands to bunker, 30
December 1947 - 1 January 1948. The buildings form part of the whal-
ing/sealing station of the Bossiere brothers, in operation 1908-31.
Alan Campbell Drury photo, Australian Antarctic Division

MV *Tottan* alongside Hobart after return from Macquarie Island,
14 April 1952
Phillip Law photo, Australian Antarctic Division

MV *Norsel*, Dumont d'Urville Station, 1955
Phillip Law photo, Australian Antarctic Division

MV *Kista Dan,* chartered from 1953-57. Used for the voyage that established Mawson Station in 1954, pictured with anchor chains on the ice, circa 1954-55

Phillip Law photo, Australian Antarctic Division

MV *Kista Dan* at 74 degrees in Horseshoe Harbour, 1954

Pen and wash illustration, with permission of the artist Fred Elliott

MV *Thala Dan*, chartered from 1957-82
Peter McLennan photo, Australian Antarctic Division

MV *Thala Dan* (left) and MV *Magga Dan* (right) anchored in
Horseshoe Harbour off Mawson Station, 8 February 1961
Geoffrey Newton photo, Australian Antarctic Division

right across the foredeck of the *Labuan*. Not following the irregular pattern of the rivets, but right through the steel so that we were very nearly going to break in half.

As a matter of fact the crew had got to the point where they were refusing to sleep up in the front quarters which flanked the tank space, because all the food, engines and warmth were in the back, at the stern.

At night they would leave their sleeping quarters, bring all their gear and sit around in the mess room or lie on the table and deck and things. They reckoned that if the ship was going to break in half, they weren't going to be in the bit that didn't have anything in it. So much for the *Labuan*!

But when *Labuan* broke down, we had urgent need of a ship to go to Macquarie. We used the *River Fitzroy*, one of the 'river ships' used for carrying iron ore for BHP. That was a pretty terrible trip as the ship was completely unsuited for the purpose.

It was a ten thousand ton ship and it was bashed around pretty heavily by the gale on the way down and on the way back. The ship suffered damage and it was a bit frightening, as it was a big ship, without the manoeuvrability of those little ships.

It wasn't any fun playing around in the unsheltered harbours around Macquarie Island and those rocky shores. But we got away with that all right.

Then we learnt that the French had chartered a little ship called the *Tottan* from the Norwegians. She was a little sea lark and had done one or two polar trips. The French used her to go down to their Antarctic station, so we sub-chartered the *Tottan* to get us out of trouble.

She was not as big as the *Norsel*, which I shall mention

in a minute. The *Tottan* was a little tin can. She was an old ship, the engines were just about worn out and they kept breaking down. But she was shaped like an ice ship and she had an ice-strengthened welded steel hull. She was strong in that sense, and could, subject to her very little power, move through pack ice. The unreliability of the engines was the main problem.

I remember one occasion down near Macquarie Island, in the middle of a gale with forty foot waves, the engines just all stopped. We turned side-on and wallowed in this dreadful gale for twenty-four hours while the engineers tried desperately, on this rolling ship, to pull the engines apart and fix things and put them back again. The crashing that went on, everything in the ship came adrift. You were rolling fifty to sixty degrees each side and it just went on hour after hour. We were scared she'd roll right over.

How big was the Tottan?

She would have been fifteen hundred tons.

So it wasn't that much smaller than the Kista?

Oh well, she must have been near nine hundred then. It depends on how you declare tonnage. What figure did you have in mind for the *Kista*?

I thought around about fifteen hundred, empty.

I suppose the most accurate tonnage to describe is displacement, which is the empty weight of a ship. Most commercial shipyards and companies don't declare the displacement, they use gross tonnage and other more useful things, that tells you how much cargo they can carry as a payload. They are not interested in how much the ship weighs. I have never known what the displacement of the *Kista* was, but say we make it eighteen hundred. In that case, I'd say *Tottan* would be about a thousand.

So she was small?

To move up from the *Tottan* to the *Kista* was to move up to luxury. I might say that in the *Tottan,* all the accommodation, except mine, was in the forecastle. It was implausible and terrible, mainly because of the motion, but secondly, the lack of heating.

The water tanks used to freeze up, causing problems with showering and drinking water. When the weather was bad, you couldn't come back across the deck. You were stuck. The crew had a terrible time.

In the *Kista Dan* it was a lot better than that. We had better saloon accommodation, and a better galley. A third of us were better off in the main part of the ship, but twelve had to bunk in the forecastle.

Again there were all these problems with the head or toilet stopping working, the shower freezing up, the terrible motion, the constant noise and the dripping of the condensation from the plates and so on.

But by and large we loved the *Kista* and got on all right with her. When I heard that another ship, the *Thala Dan* was being built, I immediately got in first and made sure we ordered the *Thala* and said to the owners we'd take her and give the *Kista* back to them for someone else to use.

They took the *Kista* back and flogged it off to the British and we went on to *Thala*. Then they built the *Magga*. It was about the same size as the *Thala*. They were sister ships, with slight differences.

A couple of years later when we needed two ships we used both the *Thala Dan* and *Magga Dan*. So I've been on expeditions with both of those ships.

The *Thala Dan* was the best of the lot. It had very commodious saloon accommodation. After all, if you are on a long voyage, most of the men sit around a lot of the time. It was even better in this respect than the later *Nella Dan*. Then the same principle applied, when we saw the new ship, *Nella*, being built we got our claws onto it well ahead of time so no one else could grab it.

Consequently we were the first to use the *Thala* and the *Nella*. While we were using the *Nella*, if we needed two ships, we used the *Nella* and the *Thala* or the *Nella* and the *Magga*. That is the way it went for years.

The only other ship I should mention is the *Norsel* that I sailed in, to go to the Weddell Sea, with the Norwegian-British-Swedish Expedition.

The *Norsel* had been a small frigate, a small naval ship that looked like a destroyer, but only half the size. It was half finished in Germany when the war ended, and this shrewd Norwegian captain bought it on the blocks. The hull had been completed and he sailed it to Norway where he completed it, finishing the decking and the fittings to suit the ship as a sealer.

This was the ship that the Norwegian-British-Swedish Expedition could see as useful and it turned out to be very good. It was ice reinforced, reasonably well powered and could perform at about the same level as the *Kista*. It was smaller than the *Kista* but bigger than the *Tottan*.

Like the *Tottan* it had a very fast roll which made it very uncomfortable. They call a ship like that stiff, as they don't flex sideways very much and it flings back into vertical very smartly. That's a very safe thing, as you know it is stable when it does that and it is unlikely to capsize. That is why the *Kista* was so frightening!

Wyatt Earp had a total roll period, swinging from one side over and back again, of four and a half seconds. Whereas the *Tottan* had a total roll period of a more reasonable eight seconds and the *Norsel* was about the same.

The *Kista* had a total roll period of fifteen seconds which was far too long. She used to hang over and you'd have to wait for it to right itself.

At the end of our frightening journey to Mawson Station in the *Kista*, we got caught in a hurricane and lost control. She did lie broadside on and nearly capsized.

During the hurricane she was already lying over thirty degrees before she started to roll and she was rolling from there. So she'd come up to thirty and roll back down again to practically eighty or ninety. She'd hang there for seconds and your heart is going 'Bang! Bang! Bang!'. You'd think she'd gone this time, she's going right round and she'd shudder and she'd come back. Absolutely terrifying!

That is the only time in my life that I have been absolutely terrified. And the terror went on for hours. There was no relief. Morale was so low, that although it was fabulous to photograph, not a single photo was taken on board that ship.

No one believed that we were going to get out of that and no one saw any point in taking photos or writing letters as the ship was going to sink. Everything on board was going

65

down, including any letters you wrote, so you weren't bothered writing home or doing any of those things.

And the captain and crew felt the same way?

Yes. The crew's morale was always worse than the expeditioners. The crew's morale was always the first to go because they hadn't the motivation of the expeditioners or their knowledge either.

Phillip, getting back to the political aspects of the decision-making involved in raising the expeditions required to open up new territory in Antarctica. Nothing was done during the war years, but as soon as the war ended I guess people got back down to business.

Yes, I think it started in 1946 with Sir Douglas Mawson hammering the Commonwealth Government and saying 'You'd better start and do something down there'.

Weight was given to his actions by the fact that the Americans conducted Exercise Operation Highjump in 1946. They took aircraft carriers and other ships right around the continent and photo mapped huge areas.

They only photographed it, they couldn't map it because they couldn't obtain astrofixes to pin the photos down, but they did photograph huge expanses of the coast. Mawson said, 'Look! Look! The Americans are doing it, if we don't get down there and start doing something we are going to lose our claim!'

So towards the end of 1946 it was agreed to do something. There were various meetings of different committees that seemed to be changing their name about every month.

Well these committees gradually got together a concept of an expedition to Macquarie Island. Mawson was keen to re-establish it as a meteorological centre. He wanted something done, to have a look near his old base at Commonwealth Bay to see whether you could put a permanent station at Cape Freshfield, which he thought was a better place. That was the general idea.

In the middle of all that, there was an urgent and secret cable from the British Government saying, 'Look, if you were going to do something in the way of expeditions you must occupy Heard Island. The Americans had always claimed that they saw Heard Island first, rather than the British.'

They'd always disputed any idea of a British claim to Heard Island and this latest work by the Americans made the British frightened that the Americans might suddenly make a claim to Heard Island. So they said, 'Get down there and stick a station on it!'.

So would it be fair to say then, that in that post-war period, latter-day imperialism was part of it.

No. It was the whole part. There was no other reason. It was straight territorial ambition. Science was just put in as something to do along the way and a smokescreen to put to the rest of the world. So that another grab or claim to territory didn't look too bad on the surface.

I know that when I was appointed Chief Scientist, and designing the program for the *Wyatt Earp*, I had great difficulty getting any real support for it. It meant stopping the ship at certain places, for marine science, on the way down and trying to get ashore to take gravity readings.

Neither the leader, Stuart Campbell nor the Captain, Karl E Oom, or any of the Committee, except Mawson, seemed to be very much interested in backing me on the sorts of things I wanted to do with the scientific section. When we got to sea, scientists were much on the outer.

So the science seemed to be peripheral to the flag-waving exercise?

Yes. Luckily the island stations were justification for the science and with Macquarie Island, which we already owned, there were not the same territorial problems. Mawson kept hammering about the meteorological value of Macquarie. Heard Island was even better because of the westerly winds which later hit Western Australia.

So would I be fair to say that Mawson was really the driving force behind Australia getting involved in Antarctica after the Second World War?

Yes. I think you could say that Mawson, almost single-handedly, started that move and was responsible for the Government setting up ANARE. Once he persuaded the Government to set up the Planning Committee and appointed Stuart Campbell as leader, then the whole thing began to roll.

Mawson's job after that was mainly as a committee member, not so much giving advice, because there wasn't much one needed to ask him about. However his prestige with the Government was so high that if the Planning Committee wanted to push the Government into doing something, then it was tremendously valuable to have

Mawson and his old Captain, JK Davis behind you. The Government took notice of them.

Did they find that somewhat essential at various points along the way?

In the first three years, we put a lot of pressure on the Government all the time, to keep doing things, which, in 1953, decided to set up a station on the Antarctic mainland. We had to get Mawson's backing again.

I should say that the man, the Minister whose enthusiasm made it possible, was Dr Herbert Evatt. But he couldn't do it on his own, just as the Minister for Science today can't get much backing for Antarctic things in Cabinet.

Evatt carried it with the force of his personality, but he couldn't persuade any of his departments to take responsibility, because there wasn't a Department of Science in those days. The nearest thing to it was CSIRO which was not a department.

The CSIRO didn't want the Antarctic program because it could see that the logistic costs would be very high and it thought the scientific product would be very low. That would distort its productivity. The science per dollar would be reduced because of the massive logistic expense that the Antarctic program involved.

Just as an aside to that, I have a feeling that with the general public, Science has never been a vote catching issue or attractive notion.

No, no! That's right. It is just now that scientists are waking up to that and deciding to set up their own lobby

69

groups and really change that picture. In the past they've considered that science justified itself and there was no need to really go and sell it. Now they're finding that that doesn't really work.

So getting back to Evatt, when he couldn't get anybody to take this on, he had to put it into his own Department, the Department of External Affairs, now the Department of Foreign Affairs and Trade.

People used to look at me and say, 'What the hell are you doing in the Department of External Affairs? That seems to be a funny place to have an expedition?' Well as it happened it actually worked out very well, because they at least understood the political sorts of things that later became the Antarctic Treaty Organisation.

It meant that they understood the political reasons for doing the work and could see that it was needed. On the logistics and science side, they didn't know anything, which enabled me to go and do it, without any interruption.

So I was able to work much more autonomously in Melbourne, than any of my successors have been able to do. After I left, the Antarctic Division swung over to the Department of Supply. It thought it knew everything about logistics. As a result, they literally wouldn't let the Antarctic Division run anything.

They moved to the Department of Science, supposing they not only knew about logistics, but also knew about science. They wanted to run everything.

The struggle to get Government support is illustrated by the fact that, in order to get them to set up an Antarctic station, I had to agree on a trade-off by which we'd close Heard Island, and thus most of the expense of setting-up an Antarctic station would be borne within budget.

I said, 'It's not much further to take a ship. If we've got to go to Heard Island, it's only another thousand miles to Mawson. You don't have to pay for new huts. I'll transfer the huts from Heard Island. I'll transfer the diesel engines, the scientific gear, the stoves and everything. So it won't cost you much!' So on that basis we obtained reluctant approval.

When I got back in March 1954, having established Mawson Station, we still didn't have approval for the expedition the following year. Do you know that I didn't get approval and money to go down to even bring those blokes home until May of that year. That meant that I had from May until December to get the enlarged Mawson Station, plus an enlarged team, plus the Heard Island one and the Macquarie Island one, which was still going, all in place.

It was a constant battle year by year until, finally towards the end of my term, we persuaded the Department to have a sort of two or three year forward look at planning, with some sort of continuity and long term agreement in place.

That was necessary because of Antarctic ship charter arrangements. We found that chartering a ship year by year was a very poor way of doing it. You were much better to charter two or three years ahead.

That gives me a fair indication of the sort of processes that went on and the sort of fights that you had to win in those early years. Let's move on to the actual difficulties and the operational logistic aspects of setting up Mawson itself.

The first thing was organisation. I had to draw up a

plan for such an expedition and I had to get the backing of the Planning Committee and then draw up Cabinet submissions to the Minister to get it through Parliament and Treasury to get the money.

Casey was our Minister and that made it easy, as Casey, at that stage had his office at the top of Collins Steet in Melbourne, not Canberra.

I used to see Casey roughly once a fortnight over several years. When he relinquished the Ministry, his successor set up his office in Canberra and after that I used to see the Minister perhaps once every six or twelve months or even eighteen months. It made a very big difference to what you could do.

Having got my plan through, the whole question arose as to where we should go. Mawson wanted to go back to the Commonwealth Bay area or Cape Freshfield, which is close by.

I wanted to go over to where we have Mawson, Mac Robertson Land, for two main reasons. First the auroral work was much more interesting over there because Mac Robertson Land is on the auroral zone, whereas Commonwealth Bay is not. Secondly, inland from there, as you know, there were also mountains and we knew they were there because of the Operation Highjump photos.

I wrote to America and got prints of all the American photos relating to Mac Robertson Land. I was able to persuade the Committee that for those reasons it was much more interesting to go over there.

I also said, 'The Norwegians have done exploration work over there that we have to cancel out. Mawson is the only one that has been around Commonwealth Bay, so we have got that part fixed. Let's go to where we've got a real job to do politically, in annulling the work of the Norwegians.'

So once again, you very astutely assessed the politics?

Yes. So Sir Douglas agreed readily enough when all these arguments were tossed around the table. It was then agreed that we would go over to Mac Robertson Land for a look and that was left to me.

I remember getting all those Operation Highjump aerial photos and laying them out on the floor and crawling around looking at them. I then got down to looking at them with a magnifying glass. It was astonishing just how few rock outcrops of any size there were. Once you leave the Vestfold Hills, you have to go right round to the islands, west of Mawson, before you see much rock.

I looked at all this and decided that the only part on the Antarctic continent that was approachable, was one little horseshoe-shaped bit of rock, but it was hidden from view by islands.

We knew it would be a bit hard to find and that was why we wanted to make sure we had aircraft. I then blessed my foresight in having bought those two Austers from the Norwegian-British-Swedish Expedition two years earlier. We had those you see. It was as the Norwegian-British-Swedish Expedition was winding up that I persuaded them to sell the aircraft to us, plus one hut.

I got them so cheaply that I was able to do that within budget and hang onto them until such time as this expedition came up.

Actually it would be very difficult to find anywhere around Antarctica, that is better than Mawson, as a rock-based station. If you go further west from Mawson you come to a fair bit of rock, although it was mostly as islands. There's not much on the coast itself.

You'd have to go right round into Emerson Bay and Enderby Land to find anything useful. Of course that is very difficult to get into, except for a few weeks of the year, because of pack ice.

So literally we were tremendously lucky to find Mawson and put the station there before anyone else did. Once the IGY came on, everyone was looking for those sort of places.

So doing it in 1954 was critical?

Yes. You see the Russians had to go to Mirny. Mirny was a very difficult place to operate as most of the huts had to be built on snow and ice, which is not satisfactory. They had a couple of little rocky outcrops on which they built the radio station for example.

In 1954 we were lucky again deciding we could build our next station at the Vestfold Hills, a place I had had a good look at. But, that raised the next political point in persuading the Government to build a second station.

It was made possible of course by the impact of the IGY. You see the establishment of Mawson had nothing to do with the IGY. However once the IGY started we were able to prove to the Government that we were well ahead of the other nations, simply because all our equipment and staff had been running down there for several years. We'd ironed out all the bugs and we were getting some real productivity, whereas most of those countries, that started in the IGY, didn't produce much in the first year.

So I said to the Government, 'Look, the Vestfold Hills is a very good place. It is one of the few places in Antarctica, apart from Mawson, with good rock to build a station on and if we don't grab it, someone else will, and anyway, we need another observation station within Australian Territory.' I said again, 'Starting small, I'll make a station with eight people.' Something like that, and that was what we did.

It was a very difficult station to establish because we couldn't find a site with suitable coastal access, despite the fact that the Vestfold Hills is hundreds of square miles of exposed rock.

It is difficult when you go along the coast and you want the criteria that we wanted. That is: you want a place where you can get a ship to within half a mile of the coast for landing purposes; a good anchorage so a vessel can hold her position, even in a gale; a beach where you can land Army DUKWs[1], boats and things; and a reasonably elevated flat rock-based area where you can build huts and things within reach of that beach; and if possible, you want fresh water.

---

1 Army DUKW (also known as DUWK). The name comes from the model-naming terminology used by General-Motors Corporation: D - designed in 1942; U - utility (amphibious); K - all wheel drive; and W - means two powered rear axles. These were wartime amphibious vehicles designed by: Rod Stephens Jr of Sparkman and Stephens Inc yacht designers; Dennis Puleston, a British sailor; and Lieutenant Frank W Speir. The DUKW weighed 7.5 tons and operated at 6 mph on water and 50 mph on land. It was about 30 feet long, 8 feet wide and 8 feet 6 inches high.

Now it turned out that I roamed up and down that coast, day and night, without sleep for about three days searching, without success. We found it very difficult to get the ship in to where we wanted it to go. Even when we did, we found it very hard to run boats, Army DUKWs and things, because of reefs and shallow water.

We also found it very hard to approach other parts of the coast for the same reason. We tried several fiords in the Vestfold Hills hoping that we could get into them and sail up and have a nice little station on the edge of the fiord. But we couldn't get across the bars they all seemed to have across their entrances. They were solid rock bars. The question of fresh water was hopeless too.

But with all those criteria, I was almost on the point of going back to Melbourne and saying 'We can't do it. We'll have to come back next year and try again.' That was the beginning of 1957.

On the last evening, absolutely exhausted, after three days of no sleep and continual searching, I decided to have a second look at a place we'd already rejected. I went back and had another look and realised that with a lot of bulldozer-work we could make a bit of a track from the beach up onto a flat area on top of a slight rise. So we decided to have a go at it.

It actually worked out very well, except for drinking water. The first year there was a big snowdrift close to the station. They had to use it for water and they got by, but the next year there wasn't so much precipitation. The Vestfold Hills is a desert area and we could see the water problem was going to be acute.

Even when we were first down there, we had to use the motorboat on the ship to lasso floating bergy bits, drag them in and chop bits out of them to melt for fresh water. Once the winter came they had enough snowfall to get by. It was in the summer months that the trouble existed.

None of the local lakes were fresh enough to get by?

Well it is a long way to go and carry water. The next year we brought down desalinating equipment, and we desalinated sea water from then on. It is funny to be in Antarctica, in a desert, with nothing to drink, isn't it?

Well the third station was interesting too. At the end of the IGY, the Americans found that they were overstretched. They were spending a lot of money in Antarctica and didn't want to go on doing it permanently.

They still wanted some permanent presence but not at the level of the IGY. So they were closing things down. Their scientists came to me and said 'Look it's a dreadful pity to close down Wilkes Station. It's such a nice station, it's on rock and it's a good place!' (Actually I had been in to explore Vincennes Bay even before the Americans got there.) So we had a certain amount of priority in that area too.

I of course wanted it. I could see a third station in that part of Antarctica, due south of Fremantle, would help pin our claim.

It would also be very valuable scientifically and solve some of our radio problems, because our direct link from Mawson to Melbourne didn't work too well. If we did have a radio link that went from Mawson to Wilkes then due north to Fremantle, Perth and then across to Melbourne it would be a much better system.

So I could see that there were advantages in taking over this base from the Americans and I persuaded the Government that this would be a good idea. There were then all sorts of comings and goings between Washington and Australia.

My agreement with the American scientists was that we would take over the station and that would be completely Australian station, run by Australians. If the Americans wanted to send scientists there we'd be glad to have them, but they would fall under our leadership and our control.

The minute it got into the hands of Casey and the Department of Foreign Affairs people, they were so frightened of the Americans, licking up to them so much, that they were just not going to stand the ground, that I had prepared for them.

They kept backing off and the minute the State Department found that we were being soft, they moved in force and began to make demands. They demanded that the station be a joint station and that there be two leaders: an American and an Australian. They had this joint leadership at some of their other stations. It just didn't work.

I fought bitterly against this and Casey was no good to me at all. He went soft on the whole deal. He was very pro-American as you know. He was not a strong man in many ways and he let me down on these major arrangements.

So for the next two years it was always listed as a joint Australian-American station. But I refused to have dual leadership. I demanded that there be an Australian leader; that he run the station; that the bulk of the people be Australian; and we'd have American visitors as collaborators, if the scientists wanted it.

That's how it finally worked out, by keeping the pressure on the Americans, we got them out completely, within three years, and it became just an Australian station.

What was the first year that you had a joint venture there?

In 1959. I should say that the IGY ended in 1958. When I went down in 1959, we had a ceremony, and I actually took over from the Americans. Oh, and there were all sorts of diplomatic problems. I wanted to pull the American flag down and put the Australian flag up and the Americans wouldn't have a bar of it.

So we had long arguments, that had to be resolved back home. Casey finally sent a message saying that, 'You have got to leave both flags up!' and that was the final answer to this joint ownership thing.

Of course the end to the Wilkes story came when the base became untenable. It was in a hollow and had become covered over with snow, which didn't quite melt off each summer. So each summer there was a greater increase in solid ice and the whole thing became unworkable. It was obvious that we had to build another station.

That is what we did and when we built the other station we called it Casey instead of Wilkes, and the last remnant of the American tie disappeared.

In what year was Wilkes finally abandoned?

I think it was 1968. Something like that, but we started building Casey in my time. I was largely responsible for choosing the site, in terms of my experience of sites and the criteria you needed and so on.

An interesting little aside about Wilkes because I managed to visit Wilkes the year I was in Antarctica. It was well out of the ice as they had had a big thaw and we were able to get down into the buildings. It was amazing to us to find that everything was basically left the way it was when they had abandoned the station. Diesel spare parts, medical supplies, food and cutlery. It was all there.

Yes, yes, wonderful. It is fascinating to see a site like that.

In 1959 the IGY generated a lot of interest internationally and the nations that had been involved decided to formalise a treaty didn't they?

Yes. The Antarctic Treaty was already drafted. Most of the leg work for drawing and finalising it occurred at the Canberra meeting of the Treaty nations, which was somewhere toward the end of 1959.

However, it took a long time for the actual signing of all the different individual papers by all the different nations after that. It wasn't all done at once as each nation had to go through their own governments. I don't think it was actually signed and in operation until the 1961 meeting at Washington. I'm not perfectly clear on these dates, but something like that.

The end of 1959 was important, it was more than the beginning. It was the putting-in-place of the basic elements of the Antarctic Treaty. Now 1959 was a year in which we were running our three stations and we were still maintaining Macquarie Island. We had only four stations altogether. We'd closed Heard, so we were running Mawson, Wilkes, Davis and Macquarie.

That would have given us a considerable amount of say with the Treaty Nations.

Yes. We were standing in very high regard at that particular time and our influence on the Treaty was quite considerable. As a matter of fact at the Washington meeting a very interesting thing happened.

The French had got off-side for some reason or another, always tending to stand on their national dignity about various things. The French ambassador had got himself out on a limb and was almost on the point of leaving the conference, as he didn't know how, with dignity, he could crawl back again. If the French had left the Treaty conference in 1961, it wouldn't have been signed.

Our Minister, Casey was at the conference, I forget whether he was chairman or what his position was, or whether he was just over there as Minister. He was a fine diplomat and took the French minister aside, spoke to him and cajoled him, smoothed him down, flattered him and persuaded him to come back off the limb and settle down. The whole thing was then finally ratified.

Were you involved in any of those negotiations?

I attended the Canberra Treaty meeting in 1959 and also the one in Washington where the thing was finally wrapped up. I was appalled at the Washington meeting to find some of the support team of the American group, trying to blackmail us, to get certain views they wanted included, and accepted.

There was a tremendous amount of detail in this Treaty and every nation had a different view. There was a lot of lobbying going on and a lot of argument outside the room itself. That was my first experience of a mighty nation using force and power to get it's way.

In what sense was it trying to get its own way?

Well I was actually told by a top American diplomat, well not so much a diplomat but advisor, as no diplomat would have said to me what he said: 'Well Phil if you Australians don't agree with us on that particular point of view, we'll have to withdraw our support from you guys!' I said, 'Look, you're not supporting us mate, I can do without you quite nicely thanks, and even if you were supporting us, I would tell you to go and jump in the lake. Don't come that stuff on us!'

I was so naïve, I had no idea those sorts of things went on and I've been very cynical ever since with what I've seen.

Big nations make pretty damn sure they get their way most of the time, by leaning with all their might, authority, resources, promises and bribery, to make sure they get what they want. I suppose that's what being big and powerful is all about.

Okay, then obviously being involved in this Treaty

blooded you on political intrigue in a sense. Did you become more politically astute as time went on?

Actually, I withdrew more and more from the politics. The Department of Foreign Affairs, External Affairs people, I think, were sensible on this. They said, 'Now Phil, if you want to maintain your credibility with the scientists, you don't want to dirty your fingers too much with the political pool. So if you're going to be a member of the Scientific Committee on Antarctic Research (SCAR) and head of the Australian delegation to SCAR and such things as that we suggest you don't attend any Treaty meetings.' So after 1961, I didn't attend any Treaty meetings.

Did you have a deputy there?

No. The diplomats ran it all. We were of course consulted on everything. Sometimes we went to a meeting, but we didn't actually attend and sit in the seats. It is interesting that more recently the director has been attending the Treaty meetings.

However looking back it's bad from the point of view that it wastes too much of his time. It is not really essential for him to be there. I imagine it's a bit of a junket for him and makes him feel important, but he's much better off in the short time he has during the year, to prepare expeditions. He'd be better off working here on the expeditions.

What I did learn is how amazingly competent the diplomats are at handling the speechifying in the chambers where all the arguments are going on. The delicacy with which they approach subjects, the way they flatter other people and the courteous nature of all the phraseology.

The very difficult art of thinking very concisely and accurately on your feet in lawyer fashion, on the run, when the pressure is on, is amazing.

I used to think I could talk off the cuff pretty well, but I couldn't have coped with that sort of stuff, not without years of practice and experience and by the time you become an ambassador, you're awfully good at this. I used to sit back in admiration and watch them handle the tricky ones.

By 1961, when the Treaty was ratified, I guess you probably had very few political battles left to fight?

Yes. By then we had established a plateau of existence which no one was challenging very much, until in 1962, there was a mini recession. Then we were in danger of being closed down by the simple fact that Cabinet had never really been interested in Antarctica.

It always needs one or two vigorous, energetic people to keep things alive in Cabinet. If I hadn't done as much public relations work through lecturing and writing to keep the public well informed on what we were doing, I am sure that in 1962 they would have closed it down.

One of my criticisms of the Antarctic Division over the last ten years is that it hasn't continued the public relations campaign. I think that is part of the reason for the lack of adequate support.

One aspect of your high profile in your day was the rapport you had with the media. You often used to get, as you have told me, people coming up and asking you for stories.

If you obliged them by giving them stories when they wanted them, every now and then when there was a crisis and some bad publicity and you'd be able to say to them 'For goodness sake don't publish that!' and they'd say 'Okay Phil, if you say so!' It is wonderful to have that sort of rapport with the media because if you don't and they get nasty, they use those stories against you. Then you're in real trouble.

In reference to the media though, something that has always been a question in my mind, is that given the general lack of public interest in Antarctica you must have had something extra to capture the imagination of the media.

Well I think we were lucky in the sense that I had eleven voyages of exploration and that meant that at the end of each of those voyages I had a story to tell. It was always a new one and it was always exciting and I had the freedom to tell it and provide the photos and go to town and give those interviews.

Now even if the blokes had a story, by the time it goes up through the official hand out track to Canberra, it is three days before it is an official news release, and no one's interested.

Early on you did have problems speaking of your own volition. You must have had to sort that out pretty quickly?

Luckily, I inherited the right to speak without going through the Minister from the first year, when Stuart

Campbell, who was a very independent fellow, set the pattern and defied the bureaucrats. So by the time I came in that was the pattern, I just stuck to it and continued to defy them.

They couldn't really sack you on those grounds. They tried time after time to knock me into line, I refused. I'd say to them, 'I'm not going to embarrass you and if I make a statement that has a political content and step out of court you can crucify me. But so long as I stick to factual detail of expedition adventures then I don't see that it is any business of yours.'

I was very, very careful over the years never to talk politics to the media, apart from policy and the politics of money for what we needed to do. That was legitimate.

So you had the formula of sticking to the adventure and nationalistic aspect of what you had to do and getting the media and public's support. To what extent do you think the talks, lectures and slide shows that you gave, added to community support?

Oh it made it. Remember I would move about Victoria when no one was doing this sort of thing to nearly the same extent. The public's knowledge was completely different lower down the order. It was the Victorian political support stemming from the Victorian public, that were being saturated by all this stuff. That was my power base.

And did that continue through till 1966?

As soon as the Department of Supply took over they just

grabbed all this power. They wouldn't let the Director say anything.

When did that occur?

A couple of years after I left, so about 1968, something like that. It was also made worse by the fact that for four and a half years after I left there was only an Acting Director of the Australian Antarctic Division. He didn't really feel he had enough strength to defy anyone. He was afraid that if he blotted his copybook he wouldn't get the director's job when it became available.

They probably held it over him anyway.

That's right.

Phillip, getting back to the setting-up of the three continental stations in Antarctica, you first set up Mawson with the aid of the Kista Dan *in the summer of 1953-1954. Can you tell us something about that?*

Yes. It might be a bit redundant, but I had better mention how we got the *Kista Dan*. We had been unable, ever since Stuart Campbell's time, to go to Antarctica. We just didn't have a ship. The *Wyatt Earp* proved unsatisfactory and both Stuart Campbell and I, at different times, had gone around the world trying to find a suitable ship without success.

But then in 1952 I learnt that a firm called J Lauritzen & Son in Copenhagen, Denmark, had built a ship for access to the east coast of Greenland, to service the lead mines that Danish firms had there.

The east coast of Greenland is notoriously difficult to approach because of pack ice so that such a ship had to be ice strengthened.

They built the ship in 1952 and in the 1952-53 summer they chartered it to a Hollywood film company which made a film called *Hell Below Zero* which starred Alan Ladd. It was a dreadful film and it was an awful waste of money. Most of the footage was shot in Switzerland or Norway and they simply wanted to get some shots of Antarctic pack ice and whales and things.

I think in the whole film there must have been only fifteen seconds or so of real Antarctic stuff. So it was an immense cost to the film people to send the ship all that way just get a few feet of film.

We got in touch with our agents, Westralian Farmers Transport Limited in London, who were the official shipping agents for the Commonwealth Government in England and they negotiated a charter with the Lauritzen people for the 1953-54 summer.

I might say that I wrote first to Lauritzen to ask him personally. It struck me that as he was using the *Kista* in the northern summer, he might not have anything for the ship to do over the northern winter. If she was going to be tied up then he might be prepared to charter her to us. He was very enthusiastic about the proposal. My guess was right. He hadn't anything booked for the ship over what was the southern summer. So he agreed.

We achieved a reasonable charter. I remember becoming a bit of a bush lawyer because I devoted many hours to analysing every clause in the charter document and making alterations and asking for changes.

This went on in fact over the next five years, until finally, the charter, which we had developed with Westralian Farmers for the use of our *Dan* ships, was used then for years by Westralian Farmers, who often rechartered Danish ships for other nationalities.

Now as far as the approval to actually set up the station was concerned, once I knew I had the possibility of chartering a ship, I was able to draw plans for the Government as a firm proposal.

However as support for a national station hitherto had not been strong, I had to think up some way of doing it on the cheap, so that it looked as though we were getting a bargain. This is when I said to the Government that if it was were prepared to set up Mawson Station, we would close down Heard Station. The total cost would be about the same.

The plan was to start off with only ten men at Mawson, whereas we had been running fourteen at Heard. We'd transfer the buildings and equipment from Heard down to Mawson and this would cut down on costs.

Mawson and JK Davis were on the Planning Committee and after much discussion and investigation it was decided to establish the base on Mac Robertson Land.

This was adjacent to the Norwegian territory known as Dronning (Queen) Maud Land, named by Amundsen.

During the years 1927-37 businessman Lars Christensen, who had a strong interest in Antarctica, either led or sponsored several expeditions to the area. Consequently they surveyed large areas of coast.

The 1936-37 expedition was led by Christensen, who was accompanied by his wife Ingrid. They brought a seaplane to undertake coastal survey work from the Weddell Sea to the Shackleton Iceshelf. Mrs Christensen joined one of the flights to become the first woman to fly over Antarctica. In all the Norwegian party took over two thousand oblique aerial photographs covering about six thousand square miles of coast from Coats Land to Enderby Land.

Previous expeditions had also mapped the coast of Mac Robertson Land and the Norwegians had assigned an alternative name, Lars Christensen Land. This was some of the work we were keen to supercede and thus cancel out any prior claim.

Mac Robertson Land, where Mawson Station was planned, was also on the auroral zone. My chief scientist and auroral expert Fred Jacka wanted to make detailed observations of our 'southern lights' the Aurora Australis.[2] This higher, southern latitude, provided a more reliable observation point.

You mentioned JK Davis. What was his history in Antarctica?

JK Davis was Mawson's Captain. He first went down to Antarctica with Ernest Shackleton's expedition in the steam yacht *Nimrod* in 1907-1909. Subsequently he accompanied Mawson firstly on the *Aurora* in 1911-1914

---

2 In an auroral zone plasma particles in the solar wind enter the earth's atmosphere and collide with electrically charged particles from the earth's magnetic field and ionise oxygen and nitrogen atoms to create light. Green light comes from oxygen and bluer and purplish-red comes from nitrogen.

and later the *Discovery* in 1929-1930. He volunteered for War service in 1915, but was taken off active duty in 1916, to command the relief expedition in the *Aurora* to rescue stranded members of the Shackleton expedition near the Ross Sea. Later hc was Commonwealth director of navigation in 1920-1949. Davis was an early member of the Planning Committee.

Who were some of the other people on this Committee?

The rest of the people on the Planning Committee were representatives of all the operating departments. The three Armed services were represented and there was Dr Fred White head of the CSIRO. The rest were all heads of the departments, not in the services, and comprised the head of the Bureau of Mineral Resources, the head of the Bureau of Meteorology, a representative of the Department of Trade, of course, as well as a person from the Department of External Affairs.

So naturally, Casey, when he was able chaired the Committee and when he wasn't able to, I chaired it. So it was a pretty high-powered committee. Oh, and there was a representative from the Australian Academy of Science and one other.

Now as to the choice of the actual place to put the station. I examined the photographs of Operation Highjump very carefully, with a magnifying glass, to see where I could find a rock outcrop that looked big enough to put a station on and which looked reasonably accessible from the sea. Over a couple thousand miles of coast, there wasn't much there.

What sort of scale were those Operation Highjump photographs?

They were trimetric photographs taken from aircraft flying at about six to eight thousand, perhaps ten thousand feet. They were not flying over them they were an oblique photograph study.

They'd apparently flown east to west and they were looking down on the coast to the left which could be seen to a distance of ten miles or so. So you didn't have any great detail, but you could see the rock and the continent and mountains inland.

There was one little horseshoe harbour that appealed particularly because as a horseshoe it granted very nice shelter if you could get a ship into it. One of the problems was that there were a lot of islands beyond it.

The question was whether it would be too shallow water for a ship. Then if a ship could get in, whether it could get through the arms of the horseshoe at the entrance, because generally when you have a horseshoe shape like that, there are shallows connecting the two heads. We didn't know until we got there if we could actually get through.

Well the story of the approach and the difficulties involved is told in my book *Antarctic Odyssey*. Essentially there were the problems of breaking through the pack ice in the beginning, then running into fast ice that is broken winter ice, extending thirty miles to the shore.

Then we had to get through that somehow to get on to the continent, because we couldn't afford to wait until the end of February, when it would break up.

Did you go straight down?

No. We had to call in at Kerguelen Island first. Actually we called at Heard Island first and picked up the huskies. Then we went on to Kerguelen.

We needed a large amount of fuel for the return voyage. It's three and a half thousand miles each way, so it would have been about ten thousand miles by the time we got back. I was able to persuade the French to put a stack of fuel at Kerguelen for us to pick up. So after Heard Island we called at Kerguelen and picked up fresh water and fuel and then went due south to Mawson.

I recall from reading your book that there were some traumatic moments even at Kerguelen.

Even getting water on board was terrible. The weather just didn't permit the water barge to operate, because it was very unstable and if the waves were bad you just couldn't use it. We'd keep launching it and having to put it up on the wharf again, launch it, then put it up on the wharf. The wind would change and we'd do it all again. Day after day went by and our time was running out. We were using up the expedition time of the French station, which was embarrassing. It was a very difficult period.

So when did you first sight the shores of the Antarctic?

It was, I suppose, early February.

There was still at least thirty miles of fast ice?

Yes there was more fast ice then than at any time until about 1968. After I left the Division, they had a year like that. However in the whole period I was there we didn't ever have that again.

In trying to get through the fast ice we had a very difficult situation. There was a hurricane, which increased the pressure on the surface of this fast ice and broke up some of it. This then caused huge areas of flat, fast ice to just press on the ship.

That's the first time I had ever been under real pressure from ice of the sort that could crush a ship. Luckily that was towards the end of the summer and a lot of the ice was rotten. It wasn't nearly as hard as it would have been in November and December.

So you were actually in the pack ice when the hurricane came?

We were in this fast ice. We were breaking our way through it and we were solidly wedged in this trough that we had cut. Suddenly it all started to grow and twist and the ship started to heel and then got lifted up with ice being pressed under it. You could see great cracks in the ice running out and you'd hear a report like a gun shot. Then a crack would open out across the fast ice and one bit would heap up and pile up against another.

There'd be a great ridge of tumbled, broken ice as the pressure crushed it all. With the grinding against the ship, the first two or three minutes were quite frightening, because we didn't know what would break first, the ice or the ship.

But once I saw the ice begin to crumble against the side of the ship and heap up and actually fall over onto the deck and heap up along the sides, I knew that the ship was stronger than the ice. Then we didn't have to worry about being crushed. Whether we'd get the ship out or whether we'd have to winter there, of course was a different matter.

This is a photograph of the Kista *all heeled over.*

Yes. That photograph was taken after we had just backed out. The heeling effect was caused by the captain, having transferred all his fuel and water from one side of the ship to the other. This made the ship heel over, to try to make some of the ice blocks, from underneath, slip out sideways.

We had finally to dig, literally dig the ship out. We had every man on board dig for two days. Finally, by the aid of ice anchors, dragging the stern, and the ship going full speed astern she was able to back off. We had previously unloaded all the anchor chain out onto the ice and all the other moveable things we could, to reduce the weight and make the ship float higher.

That had weakened the fast ice and opened up a few cracks. After that we were able to get through to Mawson and we found that we were able to push in through the entrance of Horseshoe Harbour.

It must have been a relief when you managed to get into the harbour?

Oh, it was a tremendous relief, except that when we got to the early hours of the morning another blow started. The Captain didn't position the *Kista* such that the ship's bows faced the wind and because we were side on, the wind got under the wings of our aircraft and damaged them. The men literally rebuilt one aircraft by using the bits of two, which was a fantastic performance down there.

Did the Captain anchor in Horseshoe Harbour?

No. He secured the mooring lines fore and aft onto the rock. So that was Mawson. It took us ten days or so to build enough huts to give the men something to live in.

We then went off and did some exploration in Prydz Bay, because I had learnt about the Vestfold Hills from the Norwegian exploration. It's one of only two large ice free areas on the coast in the whole of Antarctica.

The other area we visited was the Bunger Hills that Operation Highjump found. Later the Russians and Poles explored it and it was obvious that we had to do something about the Vestfold Hills.

I made this attempt in March 1954. We only had about thirty-six hours there and had to get out as we were caught by a hurricane that nearly wrecked the *Kista Dan*.

Can you elaborate a little?

Well the Captain had omitted to fill his fuel tanks with sea water to ballast the ship. Captains are always reluctant to do this, because when they get home they have got to steam them out. It all costs a lot of money for the owner, so they'd rather avoid it if they could.

However when the hurricane hit, it was obvious that the ship was unstable and floating very high in the water and very hard to manage. Then when the Captain tried to fill his tanks with sea water, all the pipes were frozen and he couldn't do it.

So we lay on our side blown completely out of control, the ship broached side-on to the wind, lying over at about thirty degrees, and rolling further so that your stable position was thirty degrees.

When you rolled one way you came up to vertical and when you rolled the other way, you went over to sixty, seventy or eighty degrees. Being unstable, when she'd go over to seventy or eighty degrees, she'd just hang there and shudder and you'd think 'Oh no, she's going to go right over'.

Before the next voyage the owners put sixty tons of extra steel in the keel just to stiffen her up, but at the time she was in a highly dangerous condition. That was a very frightening episode.

So frightening that you told me no one even took photographs. They could not see the point as they didn't think they'd come home.

No, no. No one did anything, except shudder and tremble, and I suppose some of them prayed, but morale was so low I don't think anyone thought we'd get out of that.

How did the Captain react to all this?

Well he had about seventeen hours straight on the bridge, without sleep, trying to control it.

Where were you for most of all this?

I was down lying on my bunk which meant lying almost on the wall rather than on the bunk. There was nothing to do except hope. I used to go up every now and again to check what was happening on the bridge, but I didn't like to get in the Captain's hair. He was obviously very worried and one didn't want to natter away to him. It was interesting to see. You have no steerage way when you broach. You can go forward and you can go aft, but you can't turn in any direction because the wind just bashes you back again. You are under way, but can't make way!

What about the waves?

There were not only huge waves, but a lot of pack ice and a lot of growlers and a lot of icebergs. The growlers and bits of pack ice would be lifted on these waves and bash down against the side of the hull.

I remember one bergy bit hitting the hull, it sounded like an explosion and bits of ice just disintegrated, and went hundreds of yards in every direction.

In order to avoid the icebergs he could go forward and he could go aft, but he couldn't steer. He was being blown sideways onto the icebergs, because they, being very much deeper, didn't drift as fast as the ship.

The ship was drifting quite fast relative to the icebergs, so you'd be drifting down on an iceberg and he'd put the ship full speed ahead to sort of crab round. Then there'd be another one and he'd order full speed astern and sort of crab his way round that one too.

Of course it was dark overnight and you were working on radar to try to dodge these things. The bergy bits don't show up on radar, only the bigger icebergs. It was a terrible situation.

Were there plans to abandon ship, although there'd be no point I suppose?

I was saying to someone only this morning, that we were looking at some new lifeboats and that the lifeboats were now all enclosed.

In our day the lifeboats were open and I said to someone you just prolong the agony to go and sit in a lifeboat. You'd die very fast. Particularly as, in 1954, there were no Antarctic bases, except the British, Argentinean and Chilean stations on the Peninsula.

On the other ninety per cent of the continent there was nothing. Of course there were no Antarctic ships, no icebreakers, nothing. So if anything had happened to us, there was no earthly way anyone could have come to the rescue. So much for Mawson.

Just going back a little before we go on to Davis, you spent ten days setting up the station accommodation that summer and managed to stockpile some fuel and unload your generators and so forth from Heard.

Oh, the other thing that was difficult that I forgot to mention about Mawson, was the fast ice was still solid inside Horseshoe Harbour. So the Captain had to break his way in and then moor the ship in such a way as it rested on the unbroken ice because we couldn't use any watercraft to unload.

The method of unloading was to bring sledges up on the ice beside the ship, unload the material onto the sledges, and then haul them away with the tracked vehicles known as Weasels[3].

When you got to the shoreline there was a tide crack between the ice and the shore. The Weasels couldn't travel over that so you'd couple the sledges up to a cable across the tide crack, using a Weasel or a Ferguson tractor on the shore side and drag them through and on up to camp.

What were the Army's amphibious DUKWs used for on that first voyage?

We didn't have DUKWs on that first trip. There was no space on the deck. We had two aircraft which I had bought from the Norwegian-British-Swedish Expedition in 1950 and we had one Antarctic hut we had bought and which was unloaded in Cape Town. We couldn't put it on the ship. There was no room, so I brought the hut down later, it came in handy. It's still there.

So you then pulled out and left the ten men?

Yes. They had to build a couple of store huts and over the next three or four years we gradually built up Mawson.

---

3 M29 Weasels were tracked vehicles developed by Studebaker Chrysler during World War II for use in snow or polar regions. They were powered by a Studebaker 70 hp Model 6.170 Champion engine motor and were capable of pulling sledge vehicles. Australia purchased several for Antarctic work in the early 1950s and three were shipped down in 1954 to establish Mawson Station. They had a range of 165 miles at 20 mph.

MV *Magga Dan*, 1962
Phillip Law photo, Australian Antarctic Division

MV *Nella Dan*, 1965
Phillip Law photo, Australian Antarctic Division

Phillip hoists the Australian flag and
names Mawson Station, 13 February 1954
R Thompson photo, Australian Antarctic Division

Mawson Station, 1954
Phillip Law photo, Australian Antarctic Division

Exploration party using Weasel over-snow vehicles,
inland from Mawson Station, 1955
J Bechervaise photo, Australian Antarctic Division

A dog sledging
party travelling
inland from
Mawson Station,
1954
*Bruce Stinear photo,
Australian Antarctic
Division*

Aluminium
buildings at
Mawson
Station, 1965
*Phillip Law
photo, Australian
Antarctic
Division*

Mawson Station with Mount Henderson on horizon, 1999
Wayne Papps photo, Australian Antarctic Division

First automatic weather station in Antarctica, Lewis Island,
Wilkes Land, established January 1958
Phillip Law photo, Australian Antarctic Division

One of the main problems was putting up a couple of seventy-foot radio masts. These were specially designed by Kelly and Lewis in Melbourne and made of steel in sections. It was quite a technical accomplishment for a small number of men to erect these. There is a special way of getting them up. You didn't put them up bit by bit. You erected them flat on the ground, then put a jury mast at right angles to the main structure and by hauling on the top of the jury mast you got the leverage to haul the thing into place. It was a very tricky operation.

Were there any particular activities of note that the men undertook in that first year at Mawson?

The main problem was travel. They did the first exploration inland of the immediate mountain ranges, and were able to gain first sight of the Prince Charles Mountains in the distance. That meant negotiating the crevasse region for the first ten miles inland. That was a hair-raising business and they had to flag a route and very religiously stick to the same route every time.

Then they made the first voyage along the coast in each direction. The one to the east was to try to get to Scullin Monolith. That was done in the winter months and with a group of four men, Lem Macey was in it with Bob Dovers as leader, and they nearly killed themselves. They suffered an horrific hurricane when they were on the sea ice at Scullin Monolith. You can't get up on to the shore there, it's too steep.

So they were on the sea ice in their Weasels. The force of the wind can be imagined when I tell you that a Weasel, which was on tracks and had a very low centre of gravity, was blown over by the force of the wind. So they sort of hung it out there and the ice was breaking up and they had to be tied onto rocks on the shore.

There was an incident, something like it, when that hurricane occurred, that trapped the *Kista Dan* as we were approaching Mawson. Because of the delay in getting through the fast ice, I had sent Bob Dovers with Weasels and equipment on, to try to get to the Mawson site and start work on the huts. So even if we took another five days or a week to break our way through, some work could be going on.

They'd got within about three or four miles of Mawson when this huge storm hit, the one that produced the pressure on the ship. They tied onto a rock on a little island and managed to half haul their caravan up onto the rock. Half of it was in the water and the other half on the rock.

One Weasel was some distance from shore, and the other one had broken through the ice. They were in danger of losing both Weasels and also in danger having their shelter blown away. There was nothing we could do. We were about twenty miles away at the height of the storm. We couldn't do anything.

Radio doesn't work in a blizzard of that sort. It doesn't work at all because you get static generated by the dry snow particles causing friction across the antenna which blocks all communications.

Dovers and his team were deeply depressed by this experience. The little island they were on was a dreadful place. It was littered with the corpses of Adelie penguins, because the fast ice had stopped the penguins feeding in the ocean. They were all starving. The chicks had died and it was like a mortuary. It was a terribly depressing place.

The only experience they had of it was this terrible wind and were just close enough to Mawson to get the downhill katabatic winds when the hurricane stopped.

Dovers and Macey came to me and suggested we give the whole experience away and go back to Australia and come back next year to try and find a better site.

I refused. I said 'Look we can't make our minds up because of just one set back like this. We've got to stay here for a few weeks, if necessary and feel our way into it'. When we got to Mawson we got a bit of goodwill and a bit of sunshine and morale improved.

The other thing Dovers suggested was that we should build a station as a temporary camp on the island and not attempt to get onto Mawson itself.

I wouldn't have a bar of that because I knew if we were on an island, we'd be like the French were later at Dumont d'Urville. They couldn't get on to the plateau, because they were separated once the ice broke up and there was open water between them and the land.

They would have been marooned for a whole year. All right then, moving on now to 1957 and Davis.

Well, as I said, from the Norwegian photographs, we knew it was essential that we get to the Vestfold Hills. At first it was not an immediate concern to set up a station there, it was mainly to explore the terrain. So we made some small attempt at exploration at the end of Mawson episode in 1954.

Then in 1955, after re-supplying Mawson, we went round to the Vestfold Hills and actually landed a bigger party and walked over a lot of the Vestfold Hills. We looked at the lakes, collected geological samples and did some aerial photography and so on. We still had the Auster aircraft.

Then in 1956 I didn't go there. We went round further exploring Wilkes Land. It was during 1956 we learnt of the big IGY business and we persuaded the Government to allow us to set up a second station because the Russians were going to put stations in our territory.

It became pretty clear that if we didn't get on to the Vestfold Hills, they would go there because it was ideal for a station.

Actually the Vestfold Hills and Mawson were about the only two reasonably accessible rock sites in the whole of Australian Antarctic Territory.

The Russians, after we beat them to get to the Vestfold Hills, went round near Hassall Island close to Mawson's old base, near the Shackleton Iceshelf, and put a station there. It was eighty per cent on ice which wasn't very satisfactory. They just had a few pimples of rocks sticking up and they managed to get their radio station onto a bit of rock, but most of their other structures were on ice.

Is it still there?

It's still there. A number of other stations have been buried and replaced. So in 1954-55 I got a feel for the Vestfold Hills. Then in 1957 I went down to try and put a station there.

It was the summer of 1957?

Yes. It was February 1957 and that turned out to be exceptionally difficult. I thought it'd be easy. There were great fiords running into this area as you know and I thought we'd just sail up on one of these fiords and put a station on the edge. It'd be protected and it'd be a nice anchorage for the ship and so on.

Well the first problem that we ran into was an immense crowd of icebergs just offshore from Davis, so we had to wend our way in through these. Then we found that there were numerous shelves and islands offshore and the water was pretty shallow and the ship couldn't get closer than half a mile to the coast. We explored up and down the coast in a DUKW to find a suitable place.

Now the trouble with the Vestfold Hills is that it's desert. There is very little precipitation and such snow that might blow around, or fall in winter, rapidly melts off in summer. There were no snowdrifts or ice drifts anywhere.

This meant the blokes, when necessary, had to chip bits of icefloe or get water any way they could until such time as we could get an evaporator down.

So finally we got the station set up and again with a very small party. It only had six people or something the first year. It was always small, as after that we built it up to a maximum of eight or ten.

Did you have a medical officer in that party of six?

Yes.

So you must have been running out of time that year too?

Again it was a very late business altogether. Yes. I can't remember without referring to my diary whether we set up

105

Davis before we went to the relief of Mawson or whether we conducted the Mawson relief first. It depended a bit on whether the ice broke out of Mawson first or whether the ice broke out of Prydz Bay first.

Was there much ice?

No. Prydz Bay is good because pack ice tends to break out pretty early. There is a swirl of current that goes pretty deep into Prydz Bay and up the other side past the Amery Iceshelf. That tends to carry the pack out.

Finally, the buildings that you set up in the first year, were they prefabricated?

They were prefabbed on the same design as that first radio hut and the later sleeping huts that Scott, Balleny and Shackleton put at Mawson. Very fine huts, put together like a pack of cards. They're still as good as ever. They're the best.

Were they aluminium?

Aluminium on the outside, yes. They were slabs of composite material consisting of aluminium each side and two inches of frothed up Bakelite solid insulation, or a Bakelite type compound.

The station at Davis was a delightful little station and a picturesque place to live in. You had very few men and it was very much easier to establish rapport and to live in a group. A beautiful little site looking out over this bay at all the icebergs.

I have often heard it referred to as the Riviera of the South.

Yes. Well the climate through the summer months was very good with lots of sunshine and lots of still weather. With no katabatic wind it was a much more pleasant environment than Mawson. However it lacked the inland features and it was also very difficult to go inland from there because of the crevassed glaciers of the ice sheet going up and down. Although there were no mountains going down, there was no great rise in land. What rise there was, was glaciated and heavily crevassed.

So to get off the Vestfold rock area at any point you had problems?

You really had difficulties, yes. But it is a wonderful place to wander around and explore. You've got all these lakes, some fresh water and some saltier than the Dead Sea. The very salty ones don't freeze in winter. It's strange, you see all the surrounding sea frozen and there's an unfrozen salty lake in the rock there.

The lake studies have been the most interesting of all the studies in the Vestfold Hills area.

Now, after those first few years, at Mawson and Davis, you were progressively building up facilities? Can you elaborate on that building program?

Yes, 1957 and 1958 of comprised the IGY. We were very successful, because we'd had all our experience at Heard and Macquarie islands to back us up. I think the Australian scientific program in the IGY was more successful per project than any other in Antarctica.

This was simply because our competitors were down for the first time and they had to conquer all the bugs in their equipment and all the problems with having forgotten things, for example having things break down and not having spares.

All the problems that you get in the first year of any Antarctic operation upset them all, no matter how big they were. The Americans, the Russians, they all had similar problems.

Over the next ten years they left us for dead of course, with their tremendous money and resources, but during those first two years, we more than matched them.

What sort of scientific discoveries were made in those first two years?

Well we were doing exploration. We discovered the Lambert Glacier, the biggest in the world, and we discovered the Prince Charles Mountains and visited them.

They hadn't been photographed prior to that?

No. They were too far inland for Operation Highjump or the Norwegian photographers to see from the coast. The Prince Charles Mountains and the Amery Iceshelf complex, abutting the Great Lambert Glacier make a fascinating complex.

Our research was in cosmic rays and upper atmosphere

physics, biology, geology, zoology, meteorology, geomagnetics and seismology.

So at the end of the IGY we were in very good shape, but the Americans found they were over stretched. They'd spent a lot of money in the IGY and they couldn't see themselves spending at the same rate for the next twenty or thirty years. So they started to cut back.

When they started closing stations, the scientists got upset because some of the observatories were very important. One of these stations was Wilkes, which plugged the gap between the Russian station at Mirny and the French station at Dumont d'Urville. It was a nice half-way position for meteorology, geomagnetism and seismology and so on.

The American scientists came to me and asked if it would be possible for the Australians to take the station over rather than have them just close it up. I said I'd look at it and finally persuaded our Government to do it, but it operated as a dual station for almost another two years as I explained elsewhere.

That was in 1959, wasn't it?

We took over Wilkes in about February or March of 1959. It was quite a ceremony. There was a Lieutenant Commander, USN in charge of the American expedition and I was in charge of the Australian party. We had a ceremony at Wilkes.

Now Wilkes Station was a very elaborate station, but poorly built. It was only built to last the IGY and was all plywood and of very cheap construction.

They had made two mistakes. Instead of having the huts separated, as we had at Mawson, they joined them up with corridors which meant a tremendous fire risk.

Secondly they built it in a hollow instead of picking rock higher up. They knew nothing about snow drift problems, you see, and they had built the station in a hollow. In the winter the station was covered with snow. Then as it was in a protected valley, the snow didn't melt in the summer.

Each year the permanent snow, around the huts, grew higher. Then on certain hot days, when it got above four degrees, a lot of this ice would melt. The water would get under the huts, rise up and lift the floorboards and things and then re-freeze.

The diesel from the generators would leak and get down through the floor boards and then, when the melt came, that would be distributed in the water right under the whole station. It would then soak in to the floorboards and we finished up with a frightening fire hazard where every floorboard in the station was saturated with diesel.

We trained the boys in fire drills. They were frightened stiff. We had the chaps practice these fire drills to a high degree of efficiency. If an alarm went off, say, accidentally in the middle of the night, there'd be someone at the site of the 'fire' within forty-five seconds. Everyone would be there within a minute and a half.

Properly dressed?

Not dressed at all. You didn't have to be. Because of the corridor system you were protected. One of the things we complained about, compared to Mawson, was that the fellows could live inside the station. Radio operators for example, in a single hut, never went outside. I just don't think that's the way to run a station.

Anyway we coped with the fire situation, but it became clear as two or three years went by, that we'd have to have another station. So my Deputy Director, Don Styles, and I worked on designing a new station. We moved all around that area looking for a site.

I picked a site two years before I retired and we spent the next two years building what was to be called Casey. It wasn't completed until two years after I left. So it took about four years to build altogether.

You started preliminary planning in 1964.

It was in my time that the site was chosen and the design was made. This particular design of a drift free station consisted of a building constructed on stilts, so that the wind would blow the snow out from underneath. It has been criticised in the last few years because of the fact that there was a lot vibration.

Now that is purely for one reason. We wanted a good girder construction, but we were not given enough money. The only way we could build that station with the money we had, was to use scaffold piping, like they use in building construction.

Now no matter what sort of end-laced water piping you put up as scaffolding, there is no earthly way you can keep it rigid. So when there was wind, the whole thing would vibrate, there'd be a whole lot of noise and the blokes used to complain.

But, it wasn't worse than the noise of the wind at Heard Island. There, a constant scream issues from the radio masts. So much so, that if the wind suddenly stopped, everyone would look around wondering what had happened. You got so used to this noise, that if it stopped, it was almost like a hit on the head.

The station lasted twenty years.

Yes. It was a brilliant idea. So much so that the Russians, the Japanese and other people copied it. It is the only way to keep a place drift free if you are on rock.

The only other criticism that I have heard of Casey is that it allowed people to live inside virtually all the time.

That's right. Yes.

Was Casey all prefabricated in Melbourne and then carted down to be bolted in position?

Yes. The huts were again the same as our famous prefabricated aluminium type huts, except due to costs, we couldn't afford the aluminium sheathing so we zinc coated steel for the outer and inner surfaces. It degenerated far quicker than aluminium and it was not as good as the aluminium in any sense. However the station was very comfortable.

We had separate huts, but they were all linked by this long caterpillar type tube which looked like a caterpillar. One side of it, the external side facing the wind, was made semi-circular so that it deflected the wind over and underneath the huts. The huts were all eight feet above the ground in this network of piping.

The setting up of Davis Station, when was that done?

In 1957 we established Davis; in January or February, I forget which; in order to have a second station for the IGY. The main lever, I had politically to get that done, was the fact that the Russians had established three stations in our territory or rather they had established one and were planning to put in another couple.

It was important that we should grab the Vestfold Hills area before any of the other IGY nations. It was one of the best sites in Antarctica and as you know rock sites are rare. Many other nations have to put up with stations built over ice.

What are some of the extra problems faced by putting them on ice?

The problem was one of inadequate stability. The ice is always moving in some way or other and you tend to melt down into it. That means that all your substructure gets wiggly and distorted and if the ice itself is moving sideways, as happens in many ice sheets, then you have further distortion. So all in all, it is much better to be firmly based on rock.

By announcing ahead of time that we were going to Davis or to the Vestfold Hills, prevented any other nation from making that their objective. In other words we stamped our right on it more or less that it was ours. The fact that we didn't get a station built for perhaps six or nine months later was not important.

The next problem was of course, with our small staff, our small facilities and just one ship, having to relieve

Macquarie Island and Mawson and then spare enough time to get in and do the Davis job.

Now in 1954, I'd had a look at the Vestfold Hills. After I established Mawson in 1955, I went back and had a good look at it. We found a site where there was a beach for the DUKWs to land on. The flat area, for building the station, was higher up with a fairly steep escarpment to get up on to it.

Talking to my more practical fellows, DUKW drivers and engineers, we wondered whether we could cut a roadway up through this escarpment and finally decided that we could do it with time. We only had another seven or eight days you see.

We looked up on the flat area and found it was big and extensive enough to put not only the station we wanted to build, but that we'd be able to extend it.

So we decided there was a fair chance we could do it, and at four or five o'clock in the morning, started unloading. While unloading and stacking stuff on the beach, we had every available man hacking away at this escarpment trying to produce a rough track the six wheel drive DUKWs could negotiate up on to the plateau.

What did the men use?

Picks and a bit of dynamite for some of the rock. Thirty or forty blokes all going for their lives with pick and shovel.

A real road gang!

Yes. So much for getting Davis off the ground. It was a very small station, I think there were only eight men the first year. With the sea evaporator and the operation of harpooning bits of ice, dragging them in, and cracking them up, they were able to get through with water. Since then of course, we have had more elaborate melting devices and evaporators, but water is still a major problem. So much for Davis. It wasn't a push-over. It was very difficult, much more difficult than Mawson.

Even more difficult than building Casey?

Yes. That was straightforward. Well I'd like to talk about our coastal exploration and perhaps a few words at the end about the inland exploration. The inland exploration was done of course, by the men from the stations. I wasn't present. But the plans for what they were to do each year, were drawn in Melbourne, under my direction. With consultation between my office, the Bureau of Mineral Resources and National Mapping in Canberra, we were faced with exploring something like four thousand miles of coast and one or two million square miles of territory.

Very little of the Australian Antarctic Territory had been explored, although Mawson had very thoroughly explored within a three hundred miles radius of Commonwealth Bay. In his western party, he had also explored within a couple of hundred or a hundred and fifty miles probably, of the Shackleton Iceshelf, where he had his most westerly base. He'd seen a few spots on the coast at other places, but had carried out no real exploration and no mapping at all of any consequence.

The Norwegians had done some very nice work from the Vestfold Hills, past Mawson and Enderby Land, across into Norwegian Territory. They did this by air. They had taken ships down and without landing, had managed to fly over some of this to get aerial photographs.

The trouble was that the aerial photographs were not pinned down accurately by latitude and longitude. The only fixes they had to position these photo mosaics, were the fixes they were able to take from the ship. Fixes from ship are seldom accurate to within more than half a mile at best, and if your chronometer is not the best it can be five or perhaps ten miles out.

So we also wanted to get into that area to explore it in order to nullify any claim the Norwegians might have as a result of their exploration.

There was a political element to that?

Oh yes. All exploration you can, say is political.

Given the fact that the Norwegians had their own territorial claim, do you think they were interested in laying another claim?

Well you never know. The whole claims situation was then so delicate that if other people got stuck into it, and the Russians and Americans in the IGY began to make claims, then the Norwegians might have been tempted to add to their own claims.

We also saw that the Russians, with three stations in our territory, were going to do a lot of exploring. If we didn't get in pretty fast and get some done, then they'd beat us to it and ultimately that might have been to Australia's disadvantage.

So for all those reasons, over the next few years, we

were vitally concerned with coastal exploration from the relief ships. As soon as they'd finish their jobs of relieving the stations, we'd try to squeeze out three to four weeks of time to get along the coast somewhere, probe in through the pack ice and try and make landings.

Then we'd get astro fixes and make scientific observations in geology and botany and other things. I can say that out of my twenty-eight visits to Antarctica, some of which included flights, I made eleven major voyages of exploration.

The first was in the *Wyatt Earp* of which I was not the leader, I was the Senior Scientific Officer. At that stage the only real exploration of any value was that which we conducted, the first real survey of the Balleny Islands. That was during the 1947-48 summer.

The second voyage was in the *Kista Dan* that discovered the Mawson site and established it and made our first landing on the Vestfold Hills.

The next one, we again landed in the Vestfold Hills but explored that area more thoroughly. We also went further south into Prydz Bay.

Then over the following voyages we explored a number of areas of coast. If I start in the extreme west we were first into Emerson Bay and first to land in Enderby Land. Then we were also the first to accurately photograph from the air, chart and map, the immense mountains in Enderby Land and Kemp Land.

Mawson had landed at Proclamation Island, but he hadn't been onto the mainland. He'd only been ashore for a few hours and he didn't have air photography and so on.

We gradually photographed all the coast from Enderby

Land, right back to the Vestfold Hills. That embraces Princess Elizabeth Land, Mac Robertson Land, Kemp Land and Enderby Land. There was no point doing the part immediately east of the Vestfold Hills, because the Russians were at Mirny, and Mawson had been very close to Mirny.

Was there a gentlemen's agreement about the distance you allowed between stations?

There was no gentlemen's agreement. It was a matter of practical priority. For example there were plenty of places on the coast that we hadn't explored so we weren't going to waste our time duplicating what the Russians did, particularly as they were so close. They'd probably do it before we did anyway.

And you had access to their information anyway?

Yes. We did. Once people had explored, they quickly made their information available, just so they could put a seal on their priorities. The Russians always gave us their own observations as soon as possible after they had made them. We did the same.

The next point of entry was Vincennes Bay, which was the area where in 1957 the Americans had set up Wilkes Station. We went in there the year before, in 1956, and landed on islands at the entrance of Vincennes Bay, in Vincennes Bay itself and on the mainland.

We did a number of flights, two hundred miles both east and west of Vincennes Bay. The ones to the west went right across to the Bunger Hills, an area very much like the Vestfold Hills.

Then we moved down near Adelie Land, not far from Mawson's Commonwealth Bay base. Commonwealth Bay is on the eastern side of Adelie Land, but we wanted to explore the western coastal region of Adelie Land.

When we did get in, we found that it was an ice cliff coast, with just one little island that we called Lewis Island. We established an automatic weather station on that island, which is a story in itself. It was the first automatic weather station in Antarctica.

Is it still operating?

Oh no! It ran for more than six months, which was a world record in terms of how long anything like that had run. Now the boys in the Antarctic Division have some automatic weather stations that last eighteen months or so.

The batteries were the trouble, as we had to have them charged by a wind generator. There were all sorts of things that could go wrong as the electronics weren't as good as they are now, they had valves instead of transistors or chips and so on.

Well having established Lewis Island, we then explored further to the west of that and immediately to the east. We identified the Dibble Iceberg Tongue.

Following the success of the automatic weather station on Lewis Island, we decided to put another one further along the coast. Again we found a pretty impossible sort of coastline, because of ice cliffs, no rock, fast ice and very difficult pack ice.

How high were those ice cliffs on average?

Generally thirty to sixty feet high, but there was no point in landing on ice cliffs anyway, if there was no permanent rock there, but we did find another little group of islands. We called one Chick Island, because it had lots of penguin chicks in the rookeries. We installed another weather station there.

How did you decide upon these names?

I was chairman of the Place Names Committee for Antarctica. We used to have submissions made by men from expeditions and we ourselves would put in submissions of various sorts.

I've forgotten who Lewis Island was named after, but I've got a vague idea it was named after Don Lewis, who was an American observer with our expedition.

Lewis Island was the first automatic weather station, the second one, established a year or two later, was on Chick Island.

Near it were some other islands, called Henry Islands. Those were well inside the fast ice zone, so the ship had to stand a long way off. Without helicopters, we couldn't have done anything. All the gear had to be ferried from the ship to the island by helicopter. The huts were then built and the equipment set up.

From Chick Island, we explored westward, again with the helicopter, to an area called Porpoise Bay. It has a very difficult coast and since the time we explored at Porpoise Bay I don't think anyone has approached it, or even landed there.

From there on towards Casey Station is a very difficult coast. Even the Russians haven't managed to do much with it. I think there is a great future for the Antarctic Division to get in and explore further that part of the coast.

Why was it so difficult Phillip?

The pack ice was heavy, stretching a hundred miles offshore and often with thirty or forty miles of fast ice sticking there with icebergs. It was a very messy coast where it was hard to delineate just what was the ice cliff of the coast and what was the fast ice and so on. So the coastal drawing is not terribly accurate, even from air photos.

The next major area was Oates Land and again this is a very difficult area and was one of the last areas left to explore west of the Ross Sea. You might remember that Borchgrevink had a station put in at the beginning of this century at Cape Adare.

Now the portion of coast I am referring to is from Cape Adare say, through to the west, as far as Commonwealth Bay. Cape Adare is on the corner of the entrance of the Ross Sea and Commonwealth Bay is just on the edge of Adelie Land. There are about four or five hundred miles of coast along there.

I made five attempts to complete the Oates Land work and for various reasons I kept being frustrated. There was very heavy pack, and the fact that by the time we got there it was very late in the season. If you didn't get in and out within a week, then conditions became so bad, with winter coming on, that it then became too difficult.

You might recall that it was at Oates Land where the Germans lost a ship[4] that was crushed by pressure ice. The ice sweeps around the Cape Adare corner and then, under the effect of the ice drift from east to west, piles up against protuberances. Such a feature was a big glacial tongue sticking out, a snout of some thirty or forty miles long, against which the ice packed up.

There were several other areas like that one a bit further to the west, for example, near Chick Island, called the Dalton Glacier Tongue which we had named.

The Germans were situated near this feature, and became jammed between the tongue and the ice drift. I could have warned the Germans not to go to that particular spot, as we were nearly trapped there previously.

We were trapped in the sense that we were jammed in the pack ice and began to drift towards it. It was only the fact that the currents and tides had changed that it let us get out. We were within about five miles of this glacier tongue when we got out. Otherwise we'd have been crushed too.

The German ship was brand new I think?

Yes, it was. So out of my five attempts I got ashore there, one way or another, three times. One time we were blocked off simply because on our way there everything appeared fine. The ice looked good and it looked as though we'd

---

4 *Gotland II*, on charter to the German Government and under the command of Captain Ewald Brune, was crushed by pack ice off Cape Adare on the Pennell Coast in December 1981. All hands were evacuated.

have a very successful season. Then the steward came and said he didn't have any food left. He'd misordered! I was incredibly angry. It was a stupid thing. We had to give up and just go home.

Ultimately though, we got into several places. We had a fascinating time at Oates Land. Our work remained the only work for a quite a while, until the Americans and New Zealanders carried out a series of exploits in that area. By flying up there from McMurdo Sound and putting teams in with long-range aircraft, aided by helicopters. They got some conveyancing work done on the inland parts of Oates Land, whereas we'd done the work on the coastal fringe.

You had helicopters quite often in those days?

Towards the end of my career we had helicopters, but they were very small and pretty primitive. They were not jets, so you had to wait half an hour for them to warm up their petrol engines in the morning and their operating range was only sixty miles.

If you wanted to go more than sixty miles you had to pre-position a dump of petrol and then frog hop your way out from that. You were often working at the extreme limit of your resources, because of the short-range pattern. Also you were depending absolutely on good weather to get you back. So the helicopters were pretty risky.

I should imagine petrol engines wouldn't have been as reliable either?

No, but the efficiency of our coastal exploration was very high and we developed what I called hit and run tactics.

I'd been appalled reading all the old classical expedition reports to find out just how much time was wasted.

You can't afford to waste time in Antarctica. Good weather is very rare. When it comes, you have got to utilise every second of it. Within a couple of days it'll change. The blizzards will come in and for seven, eight, ten days you'd be stuck in a ship, unable to do anything. Perhaps by then the ice would change and you'd be unable to get back and so on.

So we drilled our men very elaborately. Every man on the ship had a task and we knew, two or three days before we arrived at any spot, exactly what every man had to do. We'd keep the pressure on the captain to get in to the best possible spot.

The minute the anchor dropped, the men shot out in all directions on motorboats, Army DUKWs, helicopters and fixed wing aircraft, whatever we had.

The biologists would do penguin counts, census work and look at flora and fauna. The geologists would get to work and the geophysicists would do magnetic and gravity surveys. In the meantime the aircraft would be doing photo runs. The whole place was a hive of activity and for two whole days, you wouldn't bother about sleep, you'd just go for your life.

The minute the weather broke, it was back on the ship, and you would get the hell out and go to some place a bit further up. We used the bad weather for travelling time and to catch up on our sleep.

The exploration work carried out by the men was very extensive. They worked from the stations, with the main exploration work taking place from Mawson, because after they discovered the Prince Charles Mountains and the Lambert Glacier, they had a huge area of accessible country.

Further west were the Enderby Mountains, a great, complex set of mountains, so there was much to be done in that region. It all made me feel very happy having chosen that western site for the station instead of back near Commonwealth Bay, as Mawson initially wanted.

From Wilkes, there was literally nothing to investigate, once you had explored the coastal fringes of Vincennes Bay, because inland was just ice, so we carried out mainly glaciological studies there.

It was fortunate that to the east of Wilkes or Casey Station, there was a dome of ice about three hundred miles in diameter. It was like a little replica of the dome of Antarctica and we used it for doing floe movement and stress studies on the ice of the dome.

It is now called Law Dome and it's the only decent feature my name appears on in Antarctica. This is mainly because, as Chairman of the Place Names Committee, it was not my place to put my name on anything. By the time someone else thought that perhaps something should be named after PG Law, all the opportunities had gone. It's a bit like Mawson, who had to wait for ANARE to come and name some things after him, because in his day no one named anything after him for the same reason.

So the glaciology work was done, not only on Law Dome, which is still continuing, but on traverses of international character which went further east to join in with the French and American work. The famous traverse party measured the ice depths between Wilkes and the Russian station at Vostok, in 1961, or whenever it was.

So the work out of Wilkes or Casey has been top class and

nearly all to do with ice and biology, except of course for other scientific and the observatory work at the station.

Davis has been a strange sort of place. Access to the Plateau from Davis is quite difficult because it's all steep ice with highly crevassed glaciers with steep ice protecting it. Then there are no features inland anyway. So the work from Davis has tended to be around the interesting bare rock hills: the Vestfold Hills, the Larsemann Hills further south and several groups of islands.

They've also done some excellent work on the lakes in the Vestfold Hills, which are an extremely interesting part of the scientific research of Antarctica. Lakes don't occur very often in Antarctica, but we have these lakes and also in the Bunger Hills which are salt. Some don't freeze, so you get all sorts of strange results.

They don't even skim over in mid winter?

No. So altogether an army of men have carried out a tremendous amount of exploration since 1954.

I've heard it said by the current Federal Minister for Science and Technology, that under your leadership in those early days, a prodigous amount of work was done, with no significant funds at all.

Yes.

These days the science is becoming far more expensive to do.

Not only the science has become more expensive, but

the method of running the expeditions has changed. We had a small compact headquarters organisation of highly dedicated men, who didn't care what hours they worked. They weren't so worried much about salaries and didn't ask for overtime. They prided themselves that they were not in the ordinary stamp of public service workers. They had a tremendous esprit de corps, a sort of do it yourself attitude.

We'd say, 'Look we don't have the money to buy it.' Then out at the Tottenham store, we'd get these characters of ours who were mechanics and handymen to improvise. They showed tremendous innovative skill in knocking up things that we needed.

For example, when we only had enough money to build a couple of sleeping huts, based on the fine Mawson design of aluminium coating for the outside. Instead, our fellows used the same design, but cheaper materials and zinc cladding for the outside, allowing them to produce more for us. George Smith, the storeman and Lem Macey and Dick Thompson, were amazing fellows.

We always had one eye on the cost because we had very limited funds. The tendency today is to never really query costs. You say you want so and so and you argue like mad until you get it. Once you get it the sky's the limit. No one ever bothers to make anything themselves, you order it and you get careless in the way you handle it. There's a lot of waste and the whole attitude to expense today just frightens me.

The rebuilding of our stations has been a complete nonsense from an economic point of view. I think some of the scientific work, where they just blindly order hundreds of thousands of dollars worth of equipment with no one asking them, 'Look is it absolutely essential for you to have that?'.

The number of staff has increased dramatically. I'd like to see a comparison some day, of the headquarters staff under me, and the headquarters staff at the present time and particularly the headquarters staff in Canberra.

In my day there was no one in Canberra, except one officer in External Affairs who, as well as his other desk duties, had to look after the Antarctic desk. He was the bloke I'd see every now and then.

Now they've got policy officers and all sorts of blokes in Canberra, with a huge build up in bureaucracy. There is also the disadvantage that the Director is dragged away from his main work, to answer questions and justify things. He has to go cap in hand on every tiny thing. He can't make decisions, because they are made in Canberra by this mass of bureaucrats, most of whom don't know enough about Antarctica anyway to do it properly.

So you are saying that most of the political decisions and most of the long-term planning is made by the Department of Science and Technology, rather than by the Division itself?

I wasn't thinking so much of the political long-term decisions, because there have been very few of those. I was thinking of the everyday logistic work. You see I used to work out all the charter arrangements directly through Westralian Farmers, our agents in London.

I became a bush lawyer in the sense of thrashing round the fine points of charter arrangements, trying to squeeze

Lauritzen back to the point where we'd save a few thousand dollars each year.

So far as I know, no one worries about that now. You just apply for the charter and whatever the owner asks he gets. No one gives in because no one wants to fight a legal battle over what's involved.

Yet, ironically enough money is still the problem as there is never enough to go round.

Well, that's quite ironic isn't it. Despite the fact that the money has multiplied by a factor of five, six or eight from my time, the productivity hasn't. The productivity has gone down. They are not running any more stations than we ran, in fact at one stage we were running five. When we closed Heard Island, we went back to four. I think our productivity and scientific work was higher. Of course we also had the exploration work, which hasn't happened for ten years or so.

Is it possible to say perhaps that all the easy science has been done and what's left is becoming far more time-consuming and costly to do, or is that a simplification?

I think that's a simplification. I think physics has always been the third order in difficulty. I think the easy stuff was done in biology and there is some pretty easy work waiting to be done in marine science. Most of the marine science is pretty primitive. They haven't got up to the third order in difficulty level yet. There is a lot of work of first and second order in difficulty to do, but it is time consuming and pretty expensive because of ship time and so on. But they're only using the ship time, that we were using for exploration. The charter times are not really extended much.

Just a last comment on exploration in those early years, the early 1960s for example, apart from the major traverse to the Vostok area, was the Pole ever approached from our stations?

No. We were not interested. We could not see any point in going to the South Pole. The Japanese, from a bit further west than Mawson, did a trip to the Pole and back, mainly for prestige, but scientifically it wasn't particularly important.

We did some big ice depth traverse work south of Mawson and our exploration work south of Mawson extended to between four and five hundred miles, so that was quite big.

One thing that I want to say, is that in the period of the IGY and immediately following, we had an aircraft permanently at Mawson. We built a big hanger there and were the first to permanently keep aircraft in Antarctica all year. The Americans at McMurdo have aircraft, but used to return them to the USA and not fly over winter.

We continued to operate right through winter, flying the aircraft off the sea ice outside the hanger. Then when summer came we'd transfer them up on to the plateau and fly them off the plateau ice. It is ablated or blue polished ice and you can just land on it.

How was your braking on the ice?

Wilkes Station, 1961
Phillip Law photo, Australian Antarctic Division

Casey Station, 2007
Michael Louden .... photo, Australian Antarctic Division

Station from the radio mast, Atlas Cove, Heard Island, 1954
Jack Walsh photo, Australian Antarctic Division

Station at Atlas Cove, Heard Island, 1985
Adrian Hitchman photo, Australian Antarctic Division

Supermarine Seagull V (Walrus) unloaded from LST 3501, first ANARE expedition to Heard Island, 11-28 December 1947. This aircraft was painted yellow and was piloted by Flt Lt Malcolm D Smith RAAF. It made one trip lasting one hour taking the first aerial photos of Big Ben. The aircraft was destroyed by a hurricane at Atlas Cove, Heard Island, 21 December 1947

Alan Campbell-Drury photo, Australian Antarctic Division

Bell 47G chartered from Helicopter Utilities Pty Ltd Bankstown, NSW. From the first Antarctic sortie, 9 February 1961 at Mawson to the last flight, on 12 February 1966, these aircraft (VH-UTA, VH-UTB, VH-UTC) flew a total of 740 hours

Australian Antarctic Division

De Havilland DHC-2 Beaver near Mawson hanger after blizzard
John Bechervaise photo, Australian Antarctic Division

Auster Mk 6 purchased from NBSAE in 1950
Australian Antarctic Division

Russian Lisunov Li-2T which crashed on take off, 23 December 1968, near Mawson Station after a gust blew it off the airstrip into a crevasse. All survived
Ian Toohill photo

Well the first time I ever tried to brake on sea ice at Mawson there was this blue, polished, pebbly stuff and there was an iceberg about half a mile ahead. We were in a little Auster and we landed on skis facing this iceberg. We slowed the engine, but we kept going and going and going, with the iceberg getting closer and closer on this frictionless surface. It looked as though we'd never pull up, so the pilot put the plane into a series of ground loops. He just spun it and we went waltzing down towards this iceberg. The friction sideways on the skis is always a bit greater than lengthways so gradually slowed until we ended up maybe fifty yards from the edge of this sixty-foot berg!

Getting back to aircraft, there is an old Russian wreck inland from Mawson. Do you know the story of that?[5] Did it occur in your time?

I think it must have occurred after my time. The Russians finished up later at another station, to the west of Mawson. To fly from Mirny to west of Mawson was too far for one hop, so we used to provide refuelling facilities out at the

---

5 The accident occurred on 23 December 1968. The Russian crew had overnighted at Mawson and after refuelling were destined for Molodezhnaya Station. Taxiing up the Rumdoodle airstrip, a strong gust of wind took it off the strip and it fell into a nearby crevasse.

The Russian crew escaped from the aircraft without injury. The plane was a twin-engine Lisunov Li-2T that was sent to Antarctica for the Soviet Antarctic Program in 1958.

The pilot was dismissed from the Antarctic Program and was sent home and to Siberia and nothing further is known of him. The plane lies near the airfield and is now in the 'Mawson Ice Plateau Museum of Antarctic Aviation'.

plateau airfield. This always provided some excitement for the men to rush out there to help.

This refuelling site was about twelve miles from base. Caravans were also set up there so you could sleep overnight when fuelling them. The Russians would refuel, rest and fly on.

I had heard that they were caught in a blizzard, and when it cleared, the aircraft was blown down the plateau onto its back.

Well that makes sense. That's what happened to all our aircraft, at this little airstrip, as it was almost impossible to moor aircraft. We'd have them tied down in every possible way and they'd just shake to pieces in the one hundred and forty knot winds.

One, RAAF DC3, actually tore its moorings, because it was incorrectly attached. It was blown sideways, miles across the plateau ice and dumped over the cliff edge about thirty miles west of Mawson. It's still stuck half way up the cliff.

Now Phillip, going on to a different subject altogether. You said that at a very early age, you showed a lot of interest in Antarctica and when the opportunity presented itself to become involved, you took it with both hands. As time went on it must have taken more and more of your time. How did that affect you in your normal social life for example?

Well the first point is the act of courage needed to leave a perfectly secure, tenured job at Melbourne University and plunge in to a thing that most people said wouldn't last two years. Almost every friend I had, reckoned I was mad leaving my university job to take on the Antarctic.

The IGY of course saved that. When the IGY came, all the added political pressure came on Antarctica. From then on there was no doubt that it would be a permanent arrangement.

Did you know about the IGY?

No. I took the Antarctic job on in 1949 and the IGY had not even been mooted at that stage. Later on, yes, I found it a very big strain on what might be said to be the normally accepted way of life. I suppose because of my nature, I was prepared to accept deficiencies in some areas because of the greater payoff in others.

It was extremely difficult being away from home for six months every year. I did three or four months in Antarctica, then generally a journey overseas, with another six weeks running around Australia interviewing people for jobs and frequent trips to Canberra. I suppose it all added up to about six months over the year. Over the years my wife became more and more browned off with this situation.

Were you married when you first went to Antarctica?

Yes. I was married when I was at Melbourne University. Looking back I am surprised at the equanimity with which my wife acceded to my suggestion that I should take this new job.

She became just as obsessed with Antarctica as I was

133

and she was as deeply involved in everything we were concerned with.

The little things that you don't think of, for example I am very fond of swimming, but for a number of years I was unable to swim because I was away in summer. I resented not ever being able to see pretty girls on the beach, and never seeing my wife in her summer clothing. All you see is people in dark heavy clothes. The ebullience of the summer spirit and the flimsy things you miss out on.

I wouldn't have thought of that. It's true!

There was the problem of seasickness. I never really did get over that completely. I reached the point were I was never physically ill, but I was always squeamish and only feeling fifty per cent whilst at sea.

It was a wonderful relief to get to shore at the other end and get my energy back. You just tend to be lethargic on a ship that is constantly rolling. I think it is an illness that everyone suffers. Some people skite that they don't become seasick, but I think very few people fail to become lethargic when a ship is rolling around.

There was immense pressure of work. I didn't worry about holidays or anything, although now and again I would tag a week's holiday on to an overseas trip.

I used to work fourteen, fifteen, sixteen, eighteen hours a day, as it was an 'all-in' job. That was your job and that was your complete interest in life and everything else faded into the background.

I didn't in those days even play any tennis. In my education job I used to keep myself sane by playing tennis once a week and working that off as an unbreakable appointment. But during my Antarctic days, I didn't have that fixture; so there's nothing really I could point to as a hobby or an activity outside my job.

Perhaps your work was your hobby, so therefore you didn't need one?

I was an avid reader and I found I was getting behind in my reading. I solved that by making lists during the year, of the 'Books of the Month' and other important books that were published. I organised my librarian to buy them for Mawson and put them in a box. Then instead of putting the box in the cargo, it was put in to my cabin. I'd have two or three weeks on the way down, so I'd read one or two novels a day until I arrived, and that kept me up to date with one part of life.

You never really had time to consider a family?

No. Well my wife didn't want a family and normally perhaps I'd have argued, but being situated the way I was, I could see that having a family would be no good. If I'd had children I'm sure I'd have resigned after about five years. I'd have wanted to be more the family man and spend more time with my children. So it worked out pretty well really.

Would you like to say something about what it is about the physical environment of Antarctica? I've only been there once and I've certainly got something of it in my blood. I think that makes some people want to go back again and again?

Well first of all, it is the appeal of the wild and that's the thing that every bushwalker, mountaineer, hiker and camper understands. You have the desire to get away from the shackles of civilisation and out into pure God's earth, untrammelled by man and so on.

If you ask me 'What is it about Antarctica that the others haven't got?' I think it's the sort of difference that the alps have and that bushwalking perhaps hasn't. That's another dimension.

If you get up into the vastness of the peaks, the ice, the snow, the distances and the magnitudes and all that, it's rather a different scale from hiking through the bush here in Australia.

Antarctica is just one step on again. The vastness, the magnitude of the whole science is quite overawing as you know. The first time you ever stand on a peak and look out over the ice plateau with ice mountains all around, is an incredible experience. Then I think there is the challenge of it, the adventure.

I'm a great believer in danger adding spice to things. I believe all the best sports are dangerous. For example, I wouldn't put tennis quite in the same class as skiing and football for that reason. As much as I love tennis, it's not dangerous. You're not going to get hurt unless you get hit in the eye with the ball.

So that skydiving or skiing or the various things that kids get up to today in motorboats, and surfing and so on, all have a pretty high amount of risk. This is the challenge.

Did you do much skiing over the years?

Yes, but not in Antarctica. I tended, because I was away

in summer, to take a week off in the middle of winter. I'd got up to Mount Hotham and ski with my friends. I wasn't ever able to ski overseas because I was always down south in the northern winter, which again I resented a bit. If I had been living a normal life, I am sure I'd have done some skiing overseas.

I was disappointed to find in Antarctica that the opportunity to ski was very rare. For as you know, most of the coastal fringes are ablation zones in our territory and well, you can't ski on blue ice very well. It's no fun. You go clattering down, and your turns take twenty yards to come around because you are side slipping.

It's quite frightening. If you fall there is nothing to stop you glissading on and on and on.

If you go inland, you get beautiful powder snow, but when you were so far away from your station, you didn't like to ski in case you broke a leg or something.

I have had one or two occasions of wonderful skiing on peaks miles inland, using the helicopter to take me up and whizzing down and then getting the helicopter to take me up again. You can only afford time for a couple of runs, then you'd hear duty call and the program needing you and you'd give up and go back.

So what about your studies? You were very successful in your early life. Did you get a chance to continue that?

No. That is interesting. When I did my early studies, a Master's degree was the highest available in Australia. There was no PhD. After the war they introduced a PhD and I was in the physics department and my professor said, 'Oh, you had better enrol for the PhD'. Because we could see the world, and the way the future was going, it will need people with PhDs, not a Master's.

So I enrolled for my PhD and duly began work on classical heat experiments and switched over to cosmic rays. It was in the middle of the cosmic ray work when Antarctica came up. That was fine, because cosmic ray work formed part of the studies they continued on Heard, Macquarie and Mawson stations so I stayed involved.

It was a strange situation as I'd got to the point where I'd done enough work and published enough papers to finish my PhD, but I hadn't completed the statutory period, as a part-time research officer. I was a full-time lecturer and had to take three years over a PhD, not two.

When I'd done about two and a half years I got into this Antarctic work and that took me away from the university. In those days working in Antarctica was not acceptable as working for your PhD, although now it is. So I was given, year after year, leave from the PhD program in case, at some stage, I'd get a six months to just come and sit down there.

It wasn't a question of doing more work, it was a question of completing the formal requirements. Finally I wrote to them and said, 'Look, forget it. I can't see myself ever coming back to finish.'

In the end I finished up with two honorary doctorates, one in Applied Science and the other in Education.

I was able to continue my scientific interest in all sorts of ways. I had immense fun in the early years acting as supervisor of various biological programs. We'd have young bachelors, graduates from university, coming in and we'd set them programs.

They weren't doing a higher degree and they didn't have

a supervisor in the ordinary sense, so I'd act as supervisor and help plan their programs and criticise their work and help them get their statistics right and so on.

So that must have been very satisfying?

Yes. Then I was able to continue my cosmic ray work and develop new interests in the auroral programs that Fred Jacka was just starting. For the first seven or eight years I kept pretty well in touch. Then it got out of hand, I was only able to keep in touch with the planning of the programs in head office. There you all sit around a table and argue the toss about which are the most valuable things to be doing, whether you could get the money, what the prospects were of selling the idea, if you could get the universities to help you and so on.

I was not working in the lab any more, but I still used to do a lot of reading and do critical reviews for the young blokes, on what I thought of their papers.

They'd have appreciated that I imagine.

Well, many of the papers were pretty shallow, so you didn't need much expertise to be able to direct them onto something a bit more valuable.

In those years you virtually led all the summer supply expeditions yourself, didn't you?

Pretty well. I missed out one year when I went to England following my Norwegian-British-Swedish experience. I always felt that you couldn't lead an Antarctic expedition from a desk unless you are prepared to go down every year or at least every second year.

You just don't know how the men think or the condition of the stations. I was glad I was going down, as things could go wrong so quickly at an Antarctic station. If you get a bad OIC, the whole thing can degenerate in a year.

We had a vicious problem for two or three years when men started stealing from the stations, and bringing stuff back in the ships. If I hadn't been there, that could have gone on for years, but by being there, I was able to be pretty tough and ruthless and ironed it out after a couple of years.

The main weapon I had was to threaten to have every person examined by Customs on arrival and if anything was discovered, a prison sentence for theft would result.

I said, 'I'll give you twenty-four hours to at least clear the hold. If you have anything that is not yours, no one is going to be looking. Just walk down the hold quietly tonight and put it back.'

Next morning there was a heap twelve feet high!

CHAPTER THREE

NELLIE LAW

PHIL AND NEL LAW INTERVIEW
28 MAY 1985

(Nel's comments are printed thus in this italic script)

Ian: Could I ask you how you met?

Nel: We met at University.

Phil: It was in 1936 and I was teaching and studying. I had to take time off to do a Master's and had no money or grant or anything like they have today. So I had to use up such savings as I had, to get myself through. At the time I was on unpaid leave from the Education Department.

I had just finished my Master's degree when war broke out. I was trapped in the university for optical munitions work under the Commonwealth Manpower Authority.

Over those years, we were not wealthy enough to get married and we were more or less postponing it until we were a bit more affluent.

Then Pearl Harbour occurred and the Japanese threat was on and we thought, 'Oh well, if we are going to get married, it had to be now'. So in 1941 we were married.

I am glad you remember the date.

Were you working at the university?

No. I was a member of the university, studying. I was studying all sorts of things.

So you were married during the war and lived through that experience together. Can you remember the first time that Phillip let you know he was interested in going to Antarctica?

One day at university Professor Martin said to Phil that he was worried because he didn't know anyone to lead an expedition to the Antarctic, to Heard Island in particular.

No. It was the person who was senior scientific officer under Stuart Campbell, who was to be the Chief Scientist. So Martin mentioned that to me and I said that I would be very interested. That was how I got into it.

That first year, 1947, I was Chief Scientist and as such I went down in the *Wyatt Earp* over the 1947-48 summer. Then, after I came back, I took my cosmic ray equipment on a trip to Japan and back, to get a further latitude effect, which was the thing that we were interested in.

At the end of 1948, Stuart Campbell gave up Antarctic work and went back to his job with the Department of Civil Aviation. They then made me Director.

This was the first time Nel and I were aware that it was going to be a career for me, because in 1947 and 1948 I was only on leave from the university to do this particular job.

So initially it would have appeared to both of you to be

a transient experience? How did you come to terms with the fact it might be a long term thing Nel?

Well really, you don't know, or I didn't know, what it exactly was going to entail at first. Expeditions had to be built up and people had to be interviewed and so on and the whole thing just developed. I sort of went along with it because I didn't have much choice. In the finish it turned out quite well, because about every year, he used to go away for three or four months, and it gave me a chance to paint, something which I had always wanted to do.

I often admire your work when I come here.

Thank you.

Have you had any exhibitions or done a great deal with your painting over the years?

Yes.

Do you still paint today?

Well I try to. It's a bit busy.

Talking about painting, you had a trip to Antarctica yourself. Was that one of the main motivations for going, to be able to paint in that environment?

No. I just wanted to go and see it and of course as soon as you see it, you want to paint it, because it is so beautiful.

I think that at that stage I'd been going down for some ten years and I think Nel felt that she'd like to have a better idea of what it was all about. It's funny having a husband, who's doing a job and you never see anything of what he's doing except the headquarters work back here, which is only half the story.

Right, well in that ten-year period can you think of some of the incidents that perhaps occurred that made you aware of the risks involved in the job that your husband was doing?

Oh yes! Communications would come back home occasionally, but often they didn't come back because there were radio blackouts. In that time you would just imagine what was going on, or try to imagine what was going on. That went on for years, because there wasn't so very much communication. Not at all like there is today.

So blackouts due to sun spot activity were very common?

Yes, and just part of the story.

On one occasion we had just hit a storm and we'd informed headquarters that we were having a rough passage. Almost immediately after that, there was a very serious and long radio blackout that lasted for about seven days. So Nel was back here with no news at all. However seeing that this cessation of communications coincided with the onslaught of a major storm, she became worried that perhaps the storm had damaged the ship and that we were in trouble.

I remember that I was talking to Captain Davis in the city one day and I told him about this. He said, 'Oh, Mrs Law, the seas are terrible down there.' It didn't cheer me up at all!

I can imagine. Did that occur very often or was not typical?

Oh, that was the worst in terms of the time it took and the worry it occasioned. Radio blackouts were common and there was always that bit of worry about what was happening.

Then of course when I came back with photographs, films and reports of voyages and things, there was always the story of the incidents that had happened and the narrow escapes one had had, on one occasion or another. So over the years Nel became very aware, I think, of the sorts of hazards that were involved.

Perhaps you would have preferred not to have been told some of those stories Nel?

There was one I remember very vividly where they lost a huge aeroplane. It just fell over the side of the ice cliff and they couldn't get it back, of course. It's there today, I suppose. Another occasion on which Phil and two other people, including Ayres a famous New Zealand climber, were in a little boat going toward shore. The shore was a huge ice cliff and they said 'Oh, let's climb it and get ashore that way'. So one of the climbers, whose name eludes me, was quite inexperienced,

145

but Harry Ayres was a champion. He went first with his ice-axe. Phil, in the middle, didn't have an ice-axe but the chappy behind did, which of course was an essential thing for climbing and up they went. Harry Ayres went up quite well. They were roped up and then suddenly, the chappy behind, fell and...

I think he tripped himself with his crampon and it got caught in his trousers.

So then, immediately Harry Ayres knew by the twitch in the rope, that something had happened and immediately he dug his ice-axe in and fell onto the ice and pulled and stabilised the whole situation.

You forgot to say that Bill Bewsher was the man. Not only did he fall, but when the strain came on the rope, because I didn't have an ice-axe, I was dragged out too, so the two of us were falling and Harry Ayres held us both.

How high was the cliff?

Oh, about eighty feet, and we were about three quarters of the way up, but it went straight into the sea. That was the difficulty. If we'd fallen into the sea, there was no one around. No one would have known we were there. We'd have had ten minutes of struggling and that would have been the end.

Then when Nel went south with me, she of course came face to face with all these problems. I remember on one occasion, we were going to land on a small island because we couldn't get any closer to the continent because of the ice.

This little island lay a few miles offshore and we could reach it in a motorboat. So I called for three or four volunteers from the men. It was a lousy morning, it was blowing hard, the temperature was about fifteen degrees centigrade below. It was a miserable day and the men were not too eager to go. I finally dug up a few who were prepared to, then Nel volunteered to go, which rather shamed the fellows, who had been hanging back.

The first problem was getting into the boat, due to the sea pitching up and down the side of the ship. She went down on the rope and hung there until someone said 'Jump!'. She jumped and they caught her down below. Then we chugged in and the motorboat was moored up against a flat piece of ice.

'You have forgotten to say that there was a great audience at the time whilst all this was going on. All the men who didn't go, lined up on the deck to watch us. I think they were making bets.

When we went ashore we had our crampons, but Nel didn't. This flat shelf, we had to cross, was highly polished blue ice. It was so slippery you could only cross it by crawling on hands and knees, you couldn't stand up.

Then we went on with the work of climbing to the peak, where we built a cairn, did survey work and took magnetic observations.

As Nel didn't have crampons, she stayed below, and wandered around looking at penguins and things for ten minutes.

I was sketching. But I got so cold.

She got so cold, that she had to retreat to the motorboat and crouch over the engine in the cabin to try and thaw out.

But the hazardous part was, when we turned away from the island, to go back to the ship, we struck a rock. The rock was about fifty yards out, which frightened the hell out of us all, until we found that we hadn't torn the bottom out of the boat. If we had damaged the boat and sank, it would have been pretty desperate because we'd have all been in the water.

We were in contact with a little radio set, but as the ship was about a mile and a half away, it would still have taken half an hour or more to get a boat to rescue us.

It was interesting to have Nel involved in this as she got a good idea of what life down there was like.

Nel how did you get to be there in the first place. It must be an interesting story?

I should say that I there was no way that the Department of External Affairs would ever allow me to take my wife. They comprised mainly of very cautious, lawyer type diplomats. So I decided to just smuggle her on board and not say anything about it. Well that worked out all right on the Macquarie Island voyage, when we went down and back again.

Just a test?

Nel did some nice sketching down at Macquarie Island, everything went very well, no one found out and the news didn't leak.

148

Then we were due to go over to join the ship in Perth and sail down from there to Mawson. So the idea was that Nel would get on a liner and travel from Melbourne to Perth where she would go ashore for a few days and wait. She would then join the *Magga Dan* at Fremantle to go down to Mawson.

It so happened that we'd had dinner about a month earlier with a group of people including a woman journalist from the Melbourne *Herald*.

The other people were very close friends of ours, and we confidentially told them about us going down south. The night before the ship left Fremantle, this girl spilt the beans to the *Herald* in Melbourne.

It made front page news all over Australia, including the Press in Sydney, which came out with a front page headline saying 'Seventy-Five Men and Only One Woman!'

The Department of course was furious. I was in Perth with Nel and the ship was to sail the next day. As it happened the Minister of External Affairs was John Gorton and he happened to be staying at the same hotel in Perth.

Senator Shane Paltridge had travelled on the same liner to Perth as Nel. They met and Nel had got to know his wife and of course Nel and I both knew John Gorton and his wife.

When this news broke the night before, I was immediately rung from Canberra, by the Acting Head of the Department. The Head of the Department was overseas. The Acting Head said, 'Well the Minister's in Perth!'. We put it to him and he said 'You had better go and see the Minister and tell your story, then we'll see what happens'.

Fortunately I had arranged already for the Minister to inspect the ship the next day with his wife. We waited until we were all on board, then asked them up to our cabin for a drink. After we had had a few drinks, I told them what I had decided to, that it would be nice, if I could take my wife to Antarctica with me and would the Minister give me permission. You go from there Nel.

There was great trauma, and the Minister's wife immediately said, 'Of course she has to go! She has to go!'. She backed me. Phil and Senator Gorton went downstairs and had a conference, in which he asked Phil the dangers and what not.

I was terribly upset because I knew all my clothes and everything had been packed in the cabin. I was ready. We were ready to go and to have all my clothing and so on thrown out on the wharf, with me after it, didn't appeal at all.

This was only a few hours before the ship sailed.

Luckily Senator Gorton was very kind and didn't seem to think that it was as dangerous as people could have imagined. So he gave me permission to go.

I must say that one of the factors taken into consideration was the fact that it wasn't costing the Government anything. Nel had been invited by the shipowner to travel as a passenger free of charge. That avoided one of the complications of the whole issue. Gorton was certainly very kind, but his Department was furious that he gave permission. They were infuriated that I hadn't asked them first.

As a result of that experience can you tell me some of the highlights Nel?

Well actually, it was all a highlight because it was so new. Everything about the Antarctic is different, particularly the cold which strikes you immediately. You don't realise what cold is until suddenly you have to breathe and live it for months.

The silence was the most astonishing thing. You don't know in Melbourne or wherever you live in the city, how noisy it really is all the time, and if it isn't noisy, the birds are singing or something's happening. Children are talking. But down there, there is absolutely nothing. It is just silence and you can feel it. That was the most extraordinary thing I think.

The second extraordinary thing was, that I was very lucky in my choice of time to go, because it was a year of almost constant Aurora.

What year was that?

The Summer of 1962

I think it was 1961

Oh it was one of those. It was 1961 and it got to 1962. So I was in the Antarctic in the summer of 1962 and the Aurora was just very, very vivid, for many nights at a time. In some cases it covered the sky entirely. It was the most extraordinary thing. I have never seen anything like it and neither had anyone else.

It was a great temptation to try to paint it, but as it's moving all the time. The lights are in and out, in and out and integrating and dissolving in the sky and they're all different colours. It's the most extraordinary thing to watch, but it is terribly hard to paint.

One of the other lucky things, was that it was a voyage of exploration, so that Nel not only saw Mawson Station, but lots of other parts of Antarctica, including the unknown Oates Land, which we managed to approach.

We were invited to the French Station, I dressed up as did the others, in all our woollens and mittens and what not, and three sweaters and so on. I had my hair cut very short.We approached the station and were greeted by the French. We went up a huge climb of sheer ice to get there and coming back, we had a terrible time. It was late and it was dark, and the ice was terribly slippery and we had really great fun slipping and sliding.

When we got there though, we were introduced. They didn't quite know whether I was a man or a woman. At first they said to Phil in a cable, this was previously, 'We invite the great white snowman to come and visit us.' Phil replied, 'The great white snow chief has his squaw with him' and they thought that was a joke. So when I got there, they weren't at all sure what to do. Anyhow they eventually decided that I must be a woman and the ANARE cook. The French chef made a bouquet for me out of spun sugar in all different colours. It was the most beautiful thing. He presented it to me and I made a little speech. It was a great fun, party.

Nel was actually the first woman ever to be there.

What about women at Australian stations, you would have been the first there too.

Yes I was.

She was the first British Commonwealth woman to go
to Antarctica.

*It was a great day! One day we had a cable from the Scott
Polar Research Institute saying that they'd done a lot of
research and they decided I was the first British Commonwealth
woman to ever be down there.*

How did that make you feel?

*Well, delighted in a way, quite surprised. It was quite an
accident really, but I was delighted. Some women had been
down on whaling expeditions, but they didn't land. I landed
and explored and painted and did all sorts of things ashore and
on the voyage itself.*

*Another wonderful experience was that one day we went
in to Rumdoodle. Rumdoodle is a little base at the foot of
Mount Henderson. It was the first time I had ever been in a
helicopter and a helicopter ride for the first time, is one of the
most frightening things you can have in the Antarctic.*

*The pilot, I've forgotten his name, said 'Put you helmet
on!', but I didn't, I held it on my lap, just in case. I strapped
myself in and we shot up into the air and then shot backwards
and downwards from the back of the vessel. I thought 'Oh no,
we're going into the ocean!'. I don't know whether he did it
purposely or not or whether the wind was in that direction.
Anyway he then spun around and climbed up and got altitude
and went off. Then I thought 'Heavens!'*

*I opened my eyes to see the beautiful Antarctic before me. We
were moving with such great speed, it seemed to me, towards*

that mountain, I thought we were going to become a rock painting or something. Then suddenly he stopped and landed in a jiffy. Not having been in a helicopter before and landing on ice was another experience, really. I don't recommend it.

Was that your first sight of the vast plateau?

No. You could see it from Mawson, but it was the first inland sighting.

Can you describe to me your feeling when you first sighted the plateau when you arrived by ship?

Oh, it was early in the morning and it was very misty. Then out of the mist suddenly appeared the station. I thought, I have never seen anything so lovely. We then waited for the sun to come up.

Are there any other anecdotes that you'd like to relay about that summer voyage?

One of the other jobs we had to do, was set up an automatic weather station on a little island lying offshore. It was in a region we couldn't get to because of the fast ice. We had to transport everything from the ship, about fifteen miles across the ice, by helicopter. Nel went in. You can perhaps go on from there.

Yes, I went in holding hands. I was a bit frightened about the first expedition in a helicopter, but this one was easy and we landed on the tiny island. I brought all my paints and equipment with me, because I wanted to make a study of this. It was a very rare thing, just seeing the weather station being built and wondering whether it would work. I kneeled and sketched for oh…about two, two and a half hours and got lots of sketches and a big painting of the station itself.

You just kneeled on the ice with your sketch board?

Those were some of the difficulties we had to work with.

Working with any sort of medium in painting down there is very very difficult, as everything as you know is frozen. Even oil freezes. So I had to take some stuff with me, some eucalyptus as a non freezing sort of thing, but it didn't work because it was just too cold.

So everything had to be sketched, first in crayon or pencil, or something like that. I had that sort of thing in a pocket, a great pocket I made in my parka and I took everything out that way and worked like that.

You had to work with gloves on, I presume.

I had on three pair of gloves, but I took one pair off and I worked that way. The first pair was silk, the second light woollen, and the third was a big wrap.

So you took the mitts off and you had the fingered gloves?

Yes, so I was able then to sketch.

155

So you did all your painting in that position? You would have gotten frostbite of the knees.

No. I had so many clothes on it wasn't bad actually. There wasn't much wind. Had there been wind it would have been terrible.

You did all your sketching that way, then went inside and while it was fresh in your mind, you painted it?

Yes, I painted the whole thing.

On some occasions, she would actually paint up on the ship's bridge.

Yes, up on the bridge, on a warm day, a hot day, with the sun shining. It was quite pleasant and from the bridge I could see for miles and I did lots of painting up there.

How many paintings did you produce in that summer?

Oh Lord, a lot, I don't know.

She had an exhibition of about sixty.

I've still got sketches to work up. Any time anyone wants a painting, I just get a sketch and work it up.

How long were you away?

About three months.

And you called at each of the Australian stations?

We didn't call at Wilkes. We stood off Wilkes while they took the mail in. It was a pretty rough day.

How long did you spend at Dumont d'Urville?

Just a night.

We got there in the afternoon and left about one o'clock in the morning.

Tell me about the party you had at Mawson with the two ships in the harbour. That must have been something?

It was fun. They were just together with a little plank between them.

It was a joint party in several ways. A joint party between the crew and officers of both ships.

They were the Magga *and* Thala Dan*?*

Yes. As a joint party between the ships' companies and the shore people. The shore people under my command joined the ANARE men who were living on board the two ships, and had come down on the voyage.

The party had started off as one great party on one of the ships. I packed my men off to bed at about midnight because they had to get to work the next day.

However the crew and the officers kept playing on and we were all a bit surprised the next morning, to wake up and find that they were still partying. It seemed to carry on through most of the next day.

Then we switched over to the other ship, which then became host. So literally the party went for about two nights and a day. Our men joined in, as and when they were permitted.

I imagine there wasn't terribly much work done then? That was the first occasion that the two ships had ever been moored together.

It was the only occasion they had ever been moored together.

I remember I painted them both together the day after that. I painted them both from the shore and I think I've still got the painting. I sent one as a present to Lauritzen himself, because he was the one who invited me to go down.

He was the shipowner.

Well Nel, when you got back to Australia what sort of a press reception did you get?

It was very amusing because there was a heck of a reception, lots and lots and lots of people. They all wanted me to be interviewed. One of them, I remember saying half way through the interview, 'By the way Mrs Law, did your husband go down on this trip'?

No, it was worse. It was 'Did your husband go with you'? It was amusing for me, as normally when I'd come back, as we hit the wharf, I'd be besieged by reporters, which was an awful bind as I was always very busy. However I'd have to spare an hour for their interminable questions.

On this occasion, I was able to just sneak away quietly as no one was the slightest bit interested in me. They all wanted to interview Nel. This girl hadn't a clue about the expedition, 'Tell me Mrs Law did Mr Law go with you?'

Obviously the reporter hadn't done her homework. Well, subject to that trip, did your first-hand, intimate knowledge of the environment change your perspective on Phil's role and how important his work was?

Yes. I realised how important the work was and how important the work was to him. Also from it came the realisation that lots of wives of the men who wouldn't have my experience would, I think, like to know about it.

I was besieged with invitations to speak and I kept putting them down on a note pad. When it got to six hundred I stopped, because I wasn't going to do half of them. My voice couldn't do it.

Out of this came my idea of forming an association for the wives, the Australian Antarctic Wives' Association. It is still going today and is getting stronger and stronger. To my amazement, it just keeps growing. It now extends from Queensland to Western Australia and of course now Tasmania, Sydney and Melbourne.

It performs a very valuable function in supporting the wives, who are lonely or run into difficulties, with their husbands away. This association is very helpful and makes life more tolerable for them, by offering the support of other women.

It must give you a lot of satisfaction to know that the organisation that you started is very much alive twenty years later?

Yes.

Nel was President for a number of years and now she is Patron. The first Patron was Lady Casey. Nel became Patron when Lady Casey died.

Do they have regular functions?

Yes. They meet once a month and we have a base in town. We have speakers and pictures and chat nights and all sorts of things.

And are there times when some of the young wives with perhaps very young children need others just to simply be with?

Yes. Even though now they can just ring their husbands and even though they get regular messages through all the time, they still need to talk together and get support. Some of them of course can't cope. Some really can't cope. Most of them can, but we have had some wretched incidents in which there has been real trauma. So we put a fund aside. We have a fund to support families in difficulties and that fund has been used quite a lot.

In terms of the sort of awareness that that environment creates in a person of your experience, do you think that your first-hand experience changed you in any sense?

160

Peter Shaw (left) and Arthur Gwynn (right) look on as Phillip Law
raises the Australian flag at ANARE's first landing in the Vestfold Hills,
March 1954
R Thompson photo, Australian Antarctic Division

Main camp from Hut Hill, Macquarie Island, 24 July 1948
Peter King photo, Australian Antarctic Division

Macquarie Island Station, 1962
Donald F Styles photo, Australian Antarctic Division

Macquarie Island Station, 2007
Frederique Olivier, Australian Antarctic Division

Opening Davis Station. 'At 1600 hours on 13 January 1957, work stopped and all hands assembled around a flagpole, which had been strapped to the wall of the first hut being erected, which was the sleeping hut. Phil Law made a short speech stressing the importance of the new station in the IGY program. This was followed by a short account of the achievements of Captain John King Davis, the singing of God Save the Queen, three cheers and then back to work'

Australian Antarctic Division

Davis Station, 2007

Frederique Olivier photo, Australian Antarctic Division

The 'apples' and 'melons' at Law Base, a satellite facility to Davis Station
some 120 km along the coast to the north east, 1988
Dennis Day photo, Australian Antarctic Division

Davis Station, 2007
Frederique Olivier photo, Australian Antarctic Division

Well every experience is a widening, or it should be, a widening sort of experience. I am sure it did. Also, the experience of the great beauty of the place. Nobody talks about the beauty of Antarctica as there are always blizzards and storms and things like that. But really it is a most beautiful place. I certainly don't want it spoilt. I'm very much against anything going wrong with it.

That leads into a whole gamut of issues. Where do you stand in the current situation?

Well I'm a greeny. I'm a real greeny. Yes.

You believe in a World Park concept?

Yes.

I was just going to say, it's just as well for our marriage that Nel is able to stand alone, be self-sufficient and occupy her time with her own affairs like painting and other interests. For I was away from home almost six months of every year, for almost twenty years. That is as bad as being the wife of a sailor.

But it keeps the marriage very much alive. I must say we had many honeymoons.

I'd be away three or four months down south. Then there would be another month away, while I went around Australia interviewing men. Most years I would have a month overseas. So altogether, I was away five or six months each year.

It's a long time. Even since you've retired you've still got a considerable involvement with Antarctic issues haven't you?

Yes.

So the commitment goes on for both of you?

Yes.

Well for years I was Chairman of the Australian National Committee of Antarctic Research and that continued after I retired. I was also Chairman of the Antarctic Place Names Committee which also continued on after I retired. I am now out of both. I am busy trying to set up this Australian National Antarctic Museum.

Could you perhaps tell us a little about your dreams in that area?

It started off after I learned that the *Kista Dan* was up for sale and it struck me that it would be wonderful if we could moor it in Melbourne and use it as a museum exhibit, like the Norwegians use the *Fram* in Oslo.

Like the Japanese have done in Tokyo Bay.

The *Kista Dan* was going so cheaply, we could have got it for one hundred and fifty thousand dollars which is a throw away price. It is easily the most famous Antarctic ship of all time. I'ts more famous than the *Fram* or any of those other ships. It has done very much more.

162

Its final exploit was the Transglobe Expedition that covered both poles. We used it and the British used it for the Transglobe. It was also chartered by the Argentineans and the Canadians.

However I failed to persuade the Government to purchase it, the deadline passed and finally a Greek company bought it.

I am still pursuing the idea of the museum, which is quite independent of the idea of having a ship. It is terribly important that we have a repository in which we can store all the artefacts and memorabilia of Australia's Antarctic work.

The work that we did in the early days is already history. It's over thirty years ago. Vehicles have disappeared, aircraft have disappeared. All sorts of artefacts that we used, are no longer used. They are historical mementos. They're irreplaceable.

There are all sorts of diaries and personal effects that men have, that should be promised to a museum when they die. Otherwise they will just be lost or distributed around the country.

Quite a lot of our important memorabilia has gone to New Zealand for the Antarctic Museum in Christchurch or the Scott Polar Research Institute in Cambridge. So it's essential that if Australia is to have some sort of Antarctic heritage, that we do set up an Institute of some sort.

I hope that a decision is made before I die, because I have access to so many things that other people wouldn't know about. I'd like to be appointed honorary curator, to use my knowledge, to drag into the museum all those things from around the world, that I know exist, and could get.

Are you confident that you can achieve this final goal?

I'm not confident. I've been mucked around by the Government for five years now. Whether I can ultimately get something out of it, I don't know.

I could easily do it as a private venture. We could have bought the *Kista Dan* privately, but I believe that it's the Federal Government's job to do this. It would be much more effective if it were done on a national basis. Thus I have been deferring any approach to private sources of support, in the hope that I could persuade the Government to do something.

Very soon too, the Nella Dan[6] will be twenty-five years old and just about finishing her career. That's the one I'd like, as it is named after me.

There is always a possibility?

Yes. We might get it.

They are now very aware of the issue because of the Kista incident. Well good luck with that venture. Is there anything else that I've missed out that you would like to add?

Only the fact that when the *Nella Dan* was built and

---

6 The *Nella Dan* was driven onto rocks off Macquarie Island in December 1987. After investigating salvage she was finally taken, under her own power, to deep water off Macquarie Island, and scuttled on 24 December 1987.

launched in 1962, Nel was invited by Lauritzen, the owners, to go to Copenhagen to launch the ship, named *Nella Dan,* after Nel.

The Department of External Affairs refused permission, as they regarded it as sort of payola, bribery and corruption because I was the person who used to negotiate the charters.

I thought that was being a bit far fetched, I said 'Look, I won't go and we'll pay her fare ourselves and there will be no payola about it.' But I couldn't help feeling that perhaps there was a bit of jealousy involved. I could see no reason why my wife shouldn't go at our own expense and have the honour of launching a foreign ship. I think it would have been good for Australia to have an Australian launch a Danish exploration ship.

After all it was to be called after her, in any case and it was the custom of the firm to invite the woman, whom the ship was named after, to actually launch it. That had already been done with the *Thala* and the *Kista* and I could see no reason why it shouldn't be done with the *Nella.*

No other reasons other than bureaucratic intransigence. How did you react to that Nel?

I was disappointed. I was very disappointed.

However the ship was still named after you?

Yes. Incidentally I am sure, well fairly certain, that I'm the only one to have been down south with a hole in the heart. I didn't know, but my doctor knew about it.

My doctor examined me before I went, said I was fine and

165

said 'Don't fall out of the rigging!', so off I went. I didn't have to be examined by another doctor, because I had already been examined by my own doctor.

That's a point that didn't come up in any of the queries later on, when Phil was being interviewed by Senator Gorton. That was lucky wasn't it. Well I didn't know myself, I was quite innocent.

Did you know Phil?

I knew. The doctor and I knew, but we decided to keep it a secret. We knew it would upset her.

Phil knew, but I didn't know he knew until later. It was ten years later that I had the operation.

Ten years after Antarctica?

At least ten years, wasn't it? I waited for the right technology to come along and for the right time. The doctor just timed it and organised a very brilliant surgeon and off they went.

That certainly puts you in a very special category of Antarctic woman.

It certainly does.

Thankyou Phil and Nel Law.

CHAPTER FOUR

HIS LATER YEARS

30 March 2006

In our previous series of interviews, which stretch back over more than twenty years, we followed your life with the Antartic Division. Could we discuss your life since then?

I turned ninety-six in April this year and you have kindly agreed to put together an oral history of my life. You have already recorded my Antarctic career and we are now moving to the question of my leaving the Antarctic Division and taking up a new career in education.

The reasons that I left the Antarctic job are multiple. First of all, I believe that it is a good idea to change jobs every fifteen to twenty years, to keep your batteries recharged when you tackle a new job.

I found that, at the Antarctic Division, after nineteen years, I knew all the answers. I was working, at half throttle, let's say. I did not have to look up files for anything, as I had done it all before. I felt that I wasn't really using all my resources and I was freewheeling a fair bit of the time. I thought it would be a good time to change jobs and that change would produce a greater challenge, administratively. That was one reason.

Secondly there were aspects of Antarctic work that were beginning to pall a bit. I suffered from seasickness so it was no great novelty to go on another voyage. In fact, it was rather unpleasant, a lot of the time. Then, I found that I was away from home roughly six months out of every year. I was three or four months in Antarctica, another few weeks touring around Australia and then overseas for several weeks attending conferences.

I also found that my wife was becoming a bit sick of my absences. In general we divided our lives up pretty carefully. She was quite happy for me to be away for a fair bit of the time, while she went on with her painting.

However I was getting a bit tired of missing all of the summers too. As mentioned, I was never able to go swimming or see my wife in her summer frocks because by the time I got back to Melbourne it was the winter.

Also I began to worry a bit about the dangers. I felt that I had been extremely lucky and that it was possible my luck might run out. You can't go on being lucky all of the time. Antarctic work is dangerous. You minimise the dangers by being extremely careful, but they are always lurking because of the environmental hazards.

So in one way or another, I felt that if I could find something that suited my interests, it would be a good idea to change jobs. Obviously I began to look around the field of education. Most people look at my Antarctic career and do not understand how much educational background I have had.

For example, I was more than ten years in the Education Department as a teacher and I spent nine years lecturing and tutoring in the university. I had been on the Melbourne University Council for nineteen years and on the Council of LaTrobe University for ten. So I knew lots about tertiary education administration.

Phil, was that concurrent with your Antarctic job?

Yes.

So you had the time to do both?

Yes. I was, therefore, well qualified as a university man and I also had an interest in educational matters.

So one morning when my wife was reading the *Age* newspaper she said, 'Here's a job for you Phil'. She read out an advertisement for a job as Director of the Victoria Institute of Colleges (VIC), which was a new venture being set up in Victoria.

It was to take over the existing Technical Colleges, raise their status and educational awards from a Diploma to a Degree and Higher Degree, build new campuses and re-build the old campuses and become a model that was generated at Commonwealth level.

Thus creating a tertiary pyramid of education, which was equivalent, but different from the universities. These new colleges were to be called Colleges of Advanced Education or CAEs.

When I applied for this job, I was pretty sure I would get it because I didn't know anyone who would have the administrative experience I'd had, so much of which was in tertiary education. Sure enough, I applied for the job and was appointed.

What date were you appointed?

I was appointed to start work on the 26 April 1966. I had been in the Antarctic Division from 1947 till then, a period of nineteen years, and I was to work on, in Education, from 1966 to 1977.

That job was in a totally different environment from the one you were used to and the people you dealt with were different. How did you get the whole thing moving? How did you use your energies and your talents to start a brand new ball game?

I literally started from scratch. On 26 April 1966, I went into Melbourne, stood on a street corner and said 'I am the Director of the VIC. Now what do I do? Where do I go?'

There was no office. No headquarters. There was only an Interim Council that had been set up by the State Government, to start the thing and it used to meet once a month, to generate action. They had appointed me, but they used borrowed premises for their monthly meetings.

They did not have an office. So I didn't have anywhere to go and no one telling me what to do. So I went back to the Antarctic Division and looked up its files. Three years earlier it had moved to St Kilda Road from in the city.

I looked up the addresses of the various real estate agents and made enquiries about offices to let. Finally I chose an office in the heart of Melbourne, three rooms, on the third floor of the State Savings Bank on the corner of Elizabeth and Little Collins Streets, right in the middle of Melbourne.

The noise of the trams used to come through the windows, so I had to shut them to keep the noise out. In the summer it was unpleasant.

Luckily the previous tenant had left behind a table, a chair and a coat hanger.

I walked down to the Elizabeth Street Post Office and asked them to install a telephone. I rang the Commonwealth Employment Agency and I asked them to send around three or four typists for me to choose from. When I got a typist, I then asked her to buy a typewriter. That was the beginning.

Did you have a budget of any size at that stage?

No. I think I was just working from an interim account, being paid by the Interim Council. So if I bought something I sent a bill down to the Council and they'd pay it.

My next job was to appoint a Business Manager. Later on I was to appoint a Registrar. Once I got a Business Manager and a Registrar I had the financial and scholastic field covered.

I was lucky, as in each case, I was able to appoint first class people. For the first couple of years, the three of us were Jim Carr, who'd left his job as a stockbroker to take up the position as Business Manger with the VIC, and Ron Parry, who had a background in tertiary education and became the Registrar and later, my Deputy.

It seems to me there are a lot of parallels between when you started off the Antarctic Division and the VIC.

In each case I started off with nothing. The Antarctic Division had given me great experience in how to manage that sort of situation. It all worked out very well and it ran smoothly. I was managing an umbrella organisation covering sixteen CAE Colleges and it included the big ones, RMIT, Swinburne and others.

The Teachers' Colleges as well?

No. The Teachers' Colleges later became CAEs, but under a different system. They were not under the VIC. In order to cope with the Teachers' Colleges which had been left out of the CAE system, the State Government some years later, created the State College of Victoria. That took over I think, six existing Teachers' Colleges in Victoria.

They set up a system, very similar to my VIC system. That meant that in Victoria, there were two tertiary pyramids of Advanced Education. One was the Teachers' College lot under the State College of Victoria and the other comprised the Technological Institutes under the VIC.

A couple of years after I left the VIC, the Government merged the State Colleges and the VIC into one system. Later, under the Federal Minister for Education, Dawkins, the Dawkins Plan turned them all into universities, which I felt was a very poor move. They should have been left as a separate system.

There is another incident that occurred in the time I was with the VIC, when the State Government, under Sir Henry Bolte, wanted to set up a fourth university. He wanted to put it at Ballarat. This was consistent with a promise made to the Ballarat people, some years earlier.

So he set up a committee for the fourth university and more or less told the chairman to make sure it was called 'Ballarat'.

I was very interested in the fourth university project, but I was dead against the concept of creating a fourth university. The three existing university vice-chancellors were also against it. So the three vice-chancellors and I collaborated to try and block it. I couldn't put myself up as a nominee for the council of the fourth university, but I managed to get my Deputy, Ron Parry on it.

There were many deliberations and he managed to get enough support to oppose the plan. When this committee reported to the Government, I think seven voted in favour and he, Ron Parry, had arranged six people to oppose it.

When Premier Bolte got a report from the committee with seven for, and six against, he knew, politically it wasn't on. So he scrapped the whole idea and set up another committee.

When this other committee met, the vice-chancellors and I again tried to oppose it. We particularly rejected one plan put forward, by the Director of Education, Laurie Shears. He suggested that they should make a fourth University by taking over the six State College of Victoria institutions.

We, the vice-chancellors and I, pointed out that you couldn't make a university out of a single-discipline structure. One faculty doesn't constitute a university. So ultimately, it was decided to set up another university, but to put it in Geelong.

I was furious, because what they did was to take over the Gordon Institute of Technology, which was one of my Colleges. It was the crown in my system in many ways. It was a first class institute.

We had built a new campus for it out where Deakin now stands. In various ways it was the showpiece of the CAEs.

When they took it from me, it became Deakin University in Geelong. It had an immense struggle in the first ten years, trying to make a role for itself.

When did they set up Deakin in Geelong Phil?

It must have been between about 1970 and 1973.

You talked about the vice-chancellors and yourself. You mean the vice-chancellors of the three universities and yourself as Vice-Chancellor of the VIC?

The Act called me Vice-President, because they didn't want to call me Vice-Chancellor, as it made it look like a university. They tried to create this idea of a different structure, not a university.

So I was a vice-chancellor, but I was called a Vice-President. The President was the head of the State Electricity Commission at that time. He was like the chancellor of a university. He was the chairman of the VIC Council. He had no administrative duties.

So the president's position was an honorary appointment?

Yes.

Now the CAE, when you set it up, was the main push to provide vocational education? Is that correct?

Yes. The only way to differentiate from the universities was to stress the vocationalism. We did that very effectively

174

in all sorts of ways. For example: I wouldn't let any college set up a Degree in English. I said 'If you want to teach English, I don't object to your doing that, you have to do it in a way that is different from the Universities and the same with any language.

So if you are teaching English, you set up a course in Creative Writing and Journalism. If you are teaching a language, say Japanese, you should focus on the social structure of Japan, and the history of Japan, its jargon, the language of the everyday people and historic and financial backgrounds.'

This meant that anyone doing the Japanese course at a college will not just study Japanese literature, but what would be useful for a manager of a firm who wanted to visit Japan and talk business.

So it was more practically-based than theoretical?

Yes.

How many years did you spend with the VIC? What were your major achievements?

Twelve years, 1966-77. The first major achievement was taking the Technical Colleges out of the Education Department. When the VIC was established, the main Technical Colleges were run by the Education Department. Most of them were staffed by the Education Department. RMIT and Swinburne were not. They were semi-independent.

My first big task took about two years and that was to get them out of the Education Department, because the

Director of Education wanted to retain them. He was on the VIC Council. So he was able to oppose me. But we finally won that struggle.

Then we had to reorganise the structure because administratively, the Colleges didn't know how to perform. All their finances and all their staffing and administration had been carried out by the Education Department. We made them fully independent.

The Principal had been the chief administrator. So he was a do-it-all-yourself man. For example, the Principal at Caulfield Institute of Technology used to open all of the mail for the College. So I instituted a change that I had already carried out at Melbourne University, that was, to put two people beneath the Principal. One was the Business Manager, to look after the financial affairs, and the other a Registrar, to look after educational matters,. There was thus a principal, a vice-principal and a registrar.

The next thing we had to do, was to re-build the campuses. The technical colleges were pretty ancient. Most of them needed at least refurbishing and many of them needed to be rebuilt. There were a number of instances when we had to find new campuses for them.

In some cases we amalgamated the Teachers' College campuses. For example, when I took over the Ballarat School of Mines, within two years we amalgamated it with the Ballarat Teachers' College. We did the same with the Bendigo Institute of Technology, where we amalgamated it with the Bendigo Teachers' College.

You save money when you amalgamate two colleges, by putting them on one campus. You have one student union instead of two and one library instead of two and so on.

Later on amalgamations became more common. They amalgamated them, but left them on separate campuses. That was a mistake, because it became much more expensive. You had to create a third administrative body, to run them both and pay for all the travel involved going between them.

Who were you answerable to?

I was answerable mainly to the Commission for Advanced Education, which was a Federal body. When the VIC was set up, the financing was taken away from the State and fell to the Commonwealth. We had to apply to the Commission for our funds.

So did you get any funding from the State at all?

No. As well as this, we had to get approval for any new courses from Canberra. I found great frustration in having to answer to Canberra for academic developments. A new college, for example, would have to be approved by Canberra.

For example, when we set up the first ever computer courses in Australia, we had to get approval from Canberra. I did this by appointing a Computer Committee to organise it all. We picked the best computer brains in Australia. So when the Commission in Canberra wanted to check what we were doing, they had to set up a committee on computers. They had to use the left overs because I had put all the top people on my Committee. We had the drop on them in that way.

Everything our committee proposed almost automatically went through, because Canberra didn't dare oppose the power of the people on my Committee.

You appear to have learnt to manipulate the politicians in your time with the Antarctic Division. You put that to good use by the sound of it.

Yes. I certainly learnt how to work Canberra.

Finally we created seven new campuses. We put new campuses at Ballarat, Bendigo, Warrnambool, Geelong, Gippsland and Preston.

The new colleges we created were Bendigo, Ballarat, Warrnambool, Churchill (Gippsland), the College of the Arts and the health college, Lincoln Institute.

What sort of time did that take? What was the budget?

I suppose four years. By 1970, we were rolling along pretty well. Two interesting Colleges were the Victorian College of the Arts and Lincoln Institute.

I take great pride in those, because I conceived the idea for both. The concept developed because of minor colleges in tertiary education, that were not in the VIC, wanted to become affiliated with the VIC in order to get Commonwealth funding.

They had existed with a yearly grant from the State Government which they were never sure of getting. It was very uncertain for these little Colleges and there were several of them.

The Emily McPherson School of Domestic Economy was a good example of this. I fixed that one by amalgamating it with the RMIT. An application was made to the VIC by the National Gallery School, a very famous Arts school, that only had sixty students. I had to write to them and explain that you could not make a college out of sixty students. You can't make a college out of a single discipline and you must have a multi-disciplined school of at least two to three hundred students.

Then the Albert School of Music in Albert Street, East Melbourne, which was a very old, decadent, music school, applied and at the same time the three therapy schools applied.

The three therapy schools, were independent and separate. These were physiotherapy, speech therapy and vocational therapy. At the same time the Australian College of Nursing wanted to join and the Victorian School of Ballet also applied to join.

So I looked over the field and thought I can't take the National Gallery Arts School, but if I take painting and sculpture and then take over the Victorian Ballet School and put dance in, and create a school of drama, that would work.

There was a problem with music which I will explain in a minute. So I thought if I can get all those, that were separate, rolled into one, then we could have a school of: painting and sculpture; dance and ballet; music; and drama and TV.

Hence the Victorian College of the Arts (VCA) was born. I pushed it through. I was helped along the way, because the man, at that stage who was the Director of the National Gallery Arts School, had been Head of Art at Prahran Technical College. I had two contacts with this man, one through the Prahran system and one through the National Gallery system. We appointed him Director of the VCA.

I had already done lots of work with him. We had formed a strong combination of efficiency and were able to drive through the creation of the health education system.

When I brought together the three therapy schools and the College of Nursing, I called it the Lincoln Institute because the College of Nursing was in the Lincoln Building which was in Swanston Street. So we used the name Lincoln. That is how the name Lincoln Institute came into existence.

How many years did you spend with the VIC overall?

About eleven. All of 1966, after April. I technically finished in April 1976, but we ran on until the end of that year. So I had more or less eleven years.

What happened at that point? You didn't want to renew your contract? Perhaps you felt it was time to move on?

I didn't want to renew the contract. I had turned sixty-five, and sixty-five was the compulsory retirement age in most fields. I felt it was about time I had a rest. More so, I must say because tertiary education was becoming exceedingly complex and exceedingly difficult.

All sorts of changes were being made and the complexity was developing tremendously rapidly. I found it was a wearisome task trying to keep abreast of everything. So it was with some relief that I tendered my resignation.

I can't imagine you having done nothing for very long though Phil. How long did that last?

Well, it didn't really last very long at all. I was already heavily involved with all sorts of things, including a number of committees, Antarctic affairs and various other social things I had taken on. The big difference was that in retirement I was only doing the things I wanted to do and thus avoided the boredom of full time administration. Seventy percent is administratively boring. Only thirty per cent is constructive and challenging.

So the big advantage of retirement was that you were doing what you wanted to do, not what you had to do?

That's right.

Can we go back a little bit as I remember you told me years ago, about the difficulties that you had with student demonstrations at LaTrobe University in the early 1970s during the Vietnam War. Can you talk a little about that?

I was on both the Melbourne University and the LaTrobe University Councils. We ran into the student unrest caused by the Vietnam War. I remember that student protests were held all over the place, some of them became quite violent. I was involved in two rather violent affairs. One was violent and the other thought to be violent.

The first was at LaTrobe, where the leaders of the activists decided to imprison the Council at one of its meetings. When we were at LaTrobe having our meeting, they blocked up the exit with furniture. They heaped up furniture against the exit doors. When five o'clock came and we were due to close the meeting we couldn't get out.

We had been in there for two or three hours, so it was a major issue. We also had female members on the Council and we were stuck in the room without any access to the lavatory. It was pretty desperate.

I decided I'd try to get out, so I climbed over all the furniture and began to weave my way through it and climb down the other side. I found a passageway, so I went down this into a courtyard. I found a door and a group of activists seated outside in the courtyard.

They rushed up and pushed the door against me, so it jammed me, against the doorway. They got a table and pushed it against the door and it pushed against my ribs.

For a while, I was frightened they'd crush me, but when I found that I could stand the pressure, I began to talk and discuss, then argue, yell, scream and abuse them. They were abusing me and I was abusing them.

One bloke tried to kick my face and another attempted to hit me. I said, 'Come on let me out and I'll take on any one of you.' No one took up the challenge.

Then another member of the Council followed me. He came down and said, 'Phil, you're not going to get out, give up and come back in.' I reluctantly pulled out of the door and went back inside. They then slammed the door shut.

The fellow that had followed me, had found a way out the back and so I rang the police. It was the first time anyone had invited police onto a university campus and I said 'Don't ring the Vice-Chancellor, just come and get us out of here.'

So the police were there inside ten minutes. They pulled down the furniture and the activists fled. The police released us.

By the way, the amusing thing is that, one of the students had a camera and took a photograph of me, jammed in the door yelling at them. This was on the front page of the *Sun* the next day and did their cause harm because I had a good reputation. The activist who took the photo didn't know who I was. This damaged the student cause heavily.

Why were they taking it out on you and your Council? You had nothing to do with the War?

It was all based on nonsense. For example, they attacked us because one of the councillors was chairman of BHP. The fact that he was head of BHP, which sold steel, had nothing to do with the fact that he was on Council.

I must tell you about Melbourne University, because the activists there followed the LaTrobe action by imprisoning its Council. It so happened that I arrived late for the its Council meeting and I couldn't get in because the activists had blocked the doors.

I began talking to them. It was a happy-go-lucky sort of conversation. We were almost joking. It was not very serious. They wouldn't let me in and they wouldn't let the others out. So after we talked for a while, joking and so on, I decided there was no point in staying. I decided to go home. The Council got the police and got out. This time several activists were arrested.

I should have mentioned that following the attack on me at LaTrobe the police charged two of the ringleaders with assault. They were fined two hundred dollars each and warned-off university campuses for six months.

Being activists they turned up on the university campus the next day. The police arrested them for contempt of court and they were gaoled at the pleasure of the court. That meant they had to apologise and they weren't going to apologise. The court wasn't going to let them out until they did, so they were in jail for about three months. After that time LaTrobe University said, 'That's enough'.

Following that, Melbourne University charged the leaders in its group. When the case came to court, they had various witnesses stating what happened.

The activists called me as a witness, to prove there was no violence and to prove that it was pretty amusing in the sense that it was not violent. I had to say, 'Sure it was not violent when I was joking and laughing with you blokes and didn't do anything rough. They just stopped people coming in and out.'

It was funny to be appearing for the defence in one court case and in the other, I was the cause of the charge.

That's a fascinating story. That would have been in 1971?

Something like that. Yes.

And you were still Vice-President of the VIC. You would have been well recognised in the newspaper even from your Antarctic days. Everyone would have known you.

Yes, everyone except the photographer.

Well that backfired. Moving on, you retired at the end of 1976 and you had a rest for a while, but not for long, I imagine. What was your next project?

My next major project was preserving all the heritage material, which was as a result of my rather varied career. I became busy trying to conserve documents, papers, publications, photographs, newspaper cuttings related to education and Antarctica and other things I had been involved with.

Luckily I was helped out by a group at Melbourne University, which was part of the Faculty of Education. I have forgotten the exact title, but it was a group set up to collect money publicly and to use it to employ heritage workers and conservationists to preserve scientific documents and other material.

Are you still involved to some extent?

It has grown since my time, but they were able to arrange funds to a point where they appointed a curator

185

to preserve all my documents and other heritage material. It took about six months. So as a result this heritage group produced a publication, about half an inch thick, which is an index to the documents, letters, publications, photographs and contains my own personal biography.

Where is it housed?

It is housed in the National Library in Canberra and I still have contact with this group. I still send stuff to the National Library. I am still collecting and working on it.

Well that is one of the places where this will end up, I am sure.

Also the two volumes of photographs you gave me, they will finish up there too. During the years up until 1995 I wrote four books.

After my wife had a stroke. The main problem in continuing with my work, was the cooking of meals. A trust fund was raised to pay a cook in my home for several years, to enable me to go on with writing my books and other tasks.

Dr Kathleen Ralston wrote a two volume biography of me, which took a lot of my time and hers for that matter. So literally there are six books dealing with my career.

In addition to that, I have been a long-time member of The Royal Society of Victoria (RSV). I was President many years ago and continue my association with it in various ways.

One of the things it did was to run a special symposium

on the various things I had set up and created. One of these was the Victorian Institute of Marine Sciences.

When did you do that Phil?

That was the year after Harold Holt died, around 1967 I think. A group of Melbourne scientists decided to build a memorial for Harold Holt after his death, by creating an Institute of Marine Sciences. They failed to do it because it needed Commonwealth support, John Gorton had become Prime Minister and didn't want a memorial for Harold Holt, so he wouldn't turn on the funds.

So the group thought 'We won't call it the Harold Holt, we'll call it the Australian Institute of Marine Sciences'. They found they couldn't do that because there was already one of those in Queensland at Townsville. So they said we don't care what you call it, so being Victorians, we decided to call it the Victorian Institute of Marine Sciences. We set it up under a State Act.

Who's we?

Three or four scientists from Monash University and myself. A Council was appointed and a Director chosen. It still exists at Queenscliff. For the first ten years it was very effective. It did some very important things, but I won't go into that.

So was it set up for historical record purposes or to push the boundaries of marine science and establish new strategies?

Its function was marine research. We have an island continent with a huge coastline going from temperate waters to tropical waters. We are not doing nearly enough marine science. You can add Antarctica to that.

One of the first things it did was a study on Bass Strait and all its properties, because they found the Russians and the Taiwanese were carrying out squid fishing in Bass Strait. They knew more about Bass Strait than we did.

They were in our territorial waters?

VIMS, the Victorian Institute of Marine Sciences did a marine science research study of Bass Strait covering four or five years.

One of the things they did was internationally famous. They did a study of the currents and what would happen to any oil spill from the oil wells off the Gippsland Coast. Where would it go?

They did a computer modelling exercise, and exported this model around the world. Many parts of the world had situations where they were afraid of spilling of oil.

In fact the Americans used the Australian model from Bass Strait to model the Caribbean Sea. This shows one good example of the importance of the Bass Strait project.

Is the Institute still active?

Yes, but it was married with the Victorian Fisheries Commission. I do not know what it is called now or what the status is, but it was mainly concerned with public relations work with school children and telling them about the importance of marine science. I think children visit the Queenscliff premises and so on. I don't think there is much research being done any more.

CONCLUDING INTERVIEW, 13 JUNE 2007

When I retired, I was kept fully occupied. First of all I had the continuing job of safeguarding, collecting and distributing heritage materials, documents, papers, photographs to make sure that those important items were preserved for posterity.

Agreement was reached with the National Library of Australia in Canberra, that it would look after the material. It was my job to sort, arrange, index and package it and then send it all up to the Library.

That work is still continuing and will continue until I die, I suppose.

In the interim there have been several events, which have been consequential.

The two major events were my two landmark birthdays. My ninetieth and my ninety-fifth birthday. Each was celebrated with a large occasion.

The ninetieth was celebrated at the Melbourne Cricket Ground, where the Melbourne Cricket Club helped us a lot, by accepting bookings, planning the venue and all of the services involved.

When the day came, it was tremendously successful. A huge crowd of several hundred turned up. The great majority were my Antarctic men.

189

I remember that day well. It was put on by the ANARE Club I think. Can you talk a little bit about how you established that Club?

Yes. When I first returned from the *Wyatt Earp* expedition in 1948, I went to England and found that they had an association called the Antarctic Club. This had been set up after Scott's death, following the return of his team to England.

Over the subsequent years, anyone in Australia with any Antarctic expedition experience was encouraged to join the Antarctic Club. So I, and various members of the first *Wyatt Earp* expedition and some of the members of the Heard Island and Macquarie Island parties joined the Antarctic Club. We ran midwinter dinners here in 1949 and 1950 at the same time that the British were running theirs.

It struck me that we were now an independent country and should have our own club, patterned on the British one, but not associated with it.

So on the way back from Heard Island, I think it was in 1951, I gathered some of the men around me and we tossed around a few ideas. As a result of that, we formed the ANARE club, set up a constitution, registered ourselves as a club and decided to produce an annual journal called *Aurora* and to have a midwinter dinner.

What year did you establish that?

I think the first meetings were either in 1951 or 1952. I was the first President and then after three or four years I resigned and let someone else take over. It has been

Phillip (right) with brother Peter at Mount Hotham, 1965
Peter Law collection

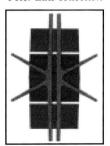

VIC logo designed by Phillip Law
Phillip Law collection

The simplified ANARE logo designed by Phillip Law, 1986
Australian Antarctic Division

Dr Phillip Law with artist Ellen Palmer Hubble
with her Archibald Prize entry for 2007
David Hubble photo

Dr Phillip Law at his 95th birthday celebration, The Melbourne Club,
20 April 2007 with Camilla van Megen and Capt William McAuley
Ian Toohill photo

Dr Phillip Law and Camilla van Megen in the Phillip Law Room,
The Royal Society of Victoria, January 2009
Bronwyn Lowden photo

Dr Phillip Garth Law
Painting and photo reproduced with the kind permission of the artist Vladimir Sobolev

going ever since, gradually expanding and very soon it established branches in every State of Australia. All these branches have a midwinter dinner, although the main one continued to be held in Melbourne. So it was the ANARE Club that put together my ninetieth birthday party.

One of the ANARE Club members was a volunteer guide taking tours around the MCG. Another man was on the staff at the MCC setting up its museum. So we had two people at the MCG ready to push it along.

It turned out to be a very fine occasion, because people came before lunch and stayed on. At the same time, they were able to look around the MCG, which was an added attraction. That was in 2002.

I recall you saying about that time that once you pass ninety, every year is a bonus. What do you think about the five bonus years you've had so far?

I never promise myself anything in the future. You never know, when you get to my age, what's going to strike you down. I am living day to day you might say, hoping that I will get to the next birthday, but not being at all surprised if I don't.

My ninety-fifth was this year [2007]. It was organised by The Royal Society of Victoria, because this event occurred during the International Polar Year (2007-2008).

It is fifty years since the IGY which was in 1957 and The Royal Society of Victoria wants to organise an expedition for school students.

It will be called RSV INTREPID (The Royal Society of Victoria's International Research Expedition Polar Inter-Disciplinary). If they get the money they are going to charter a Class I Russian icebreaker and take one hundred students, accompanied by one hundred energetic postgraduate and post-doctoral scientist members.

They will land at various points in the Australian Antarctic Territory and will send back daily reports for TV and radio. Their stories will also be distributed to schools around Australia.

The Royal Society of Victoria seized the opportunity of celebrating my birthday, to give publicity for its work in the Fourth International Polar Year. It was a dual function for the Society, but a major one for the ANARE Club and me.

This time it was a dinner, not a lunch. The Royal Society's Hall was not big enough, so they arranged with the Melbourne Club for the dinner to be held there. The Melbourne Club is able to take one hundred and eighty people for dinner and so they took bookings, and they closed the bookings when they filled up.

On the evening there were about one hundred and seventy people present, including seventeen Polar Medallists from ANARE.

It was something like an ANARE Club dinner, so it was a dual celebration. Some of the men who turned up hadn't met up for fifty years, making it a very exciting evening for the ANARE people as well as for me.

I chose five men to speak about different aspects of my career. I had indicated the sorts of things that they might talk about and made sure they didn't duplicate anything. It turned out very well.

Each one of them gave a talk, which was highly appreciated, and everyone agreed that the speeches were first class. Altogether it was a fine evening with The Royal

Society of Victoria publishing the speeches and an account of the evening and its *Proceedings* to commemorate it. There will thus be a permanent record of the event.

And there was a rather special video on the night too, which was put together for your benefit and which was very moving.

The video biography was shown while the various things were happening during the dinner.

It was an excellent night. You must have felt very proud. Is that the word? How did you feel on that night Phil?

I think 'proud' is the right word. I felt greatly honoured of course to have such an occasion.

There were a couple of other things that happened in my ninety-fifth year. A woman [Ellen Palmer Hubble] who decided she wanted to paint my portrait for the Archibald Prize[7] [2007][8]. She duly painted me and produced a very good likeness but it wasn't chosen for the Archibald which is not surprising, because it was conventional and the Archibald is not so conventional.

Recently I had another offer from a visiting Russian painter [Vladimir Sobolev in 2007] of high repute. He was visiting Australia to paint an ex-Premier of Victoria Steve

---

7 This was painted during 2006-07 with Dr Law in the robes of a Doctor of Applied Science from the University of Melbourne.

8 An earlier Archibald Prize entry, by Sally Robinson in 2002, showed Dr Law in a polar theme.

Bracks. He was also commissioned by The Royal Society of Victoria to paint me. The portrait now hangs in The Royal Society.

You also had a superb bronze done of yourself some years ago. When and how was that done?

A Vietnamese refugee [Le Thanh Nhon] sculpted that bust [in 1978] just after the Vietnam War. Remember the boat people? Well, this man was a famous artist in Vietnam, and was one of the refugees who landed in Australia, with no means of support. Until he could get some painting and sculpture commissions he had trouble keeping himself alive.

I was playing tennis with a Vietnamese chap at the time. He suggested contacting the Caulfield Institute of Technology to commission this man to do a bust of me to put in the Phillip Law Building. This is an engineering building, in the then Caulfield Institute of Technology, which is now a part of Monash University. So, I sat for this sculptor and he produced a superb bust of me. It is a beautiful piece of work and is a very strong product.

I think it depicts your personality perfectly.

I got them to cast an extra one for me personally. The original one is still in the Caulfield Institute, Monash University Campus. Mine was stolen by a drug addict who wanted to get a quick return for money for his drugs.

He thought he would flog it off to a foundry. So he looked around the foundries in Melbourne to melt it down for bronze and pay him for the weight. But, he went to the wrong foundry, he went to the one that made it! Of course they called the police, so they got the sculpture back.

Rather than keep it in the house then, I gave it to the museum to look after. Finally I decided to put it somewhere more permanent and I donated it to the Antarctic Division in Kingston, Tasmania. I am going down to Tasmania for a mid-winter dinner in a few weeks time. I'll then go and have a look at it.

Phil, you are still involved in Antarctic affairs and you take a keen interest. Where do you see Australia's involvement in Antarctica going in the next decade?

It is impossible to predict what this International Polar Year will produce. The International Geophysical Year, 1957-58 was fantastic in the results it generated. We can't expect anything like that, but one can expect huge progress in two fields of Antarctic research.

One is on global warming. Ice drilling will enable us to look at the layering of ice and track back the various climatic changes over the past thousands and even hundreds of thousands of years.

The other interesting area that is developing is that of Antarctic marine science because we really know very little about the Antarctic Ocean. They know about the seals, the penguins and the birds. However marine cartography, marine oceanography and marine life, krill and other small microscopic elements, all that is going to develop very slowly.

Even in the last few months, there have been some interesting ice cores brought up from the bottom of the Amery Iceshelf. These show unexpectedly, that the same sort of mix of tiny microscopic creatures exist under the ice shelves as in the open ocean. That's quite unexpected. They thought they'd be different because they were living with limited light.

When the Antarctic Treaty runs out and has to be renewed what are the prospects for Antarctica?

The Antarctic Treaty is vital to our future. I think it is going to play a very prominent role because the claims to Antarctic Territory are dead and buried. We persist with them because, at the moment, there is no replacement and we don't want to leave a vacuum. The minute some regulatory governmental set-up is organised for Antarctica, I am sure no country will give away its claims.

The obvious thing to do is NOT to let the United Nations muck around with it, but to let the Treaty nations operate it. For to be a voting member of the Treaty, you have to have a station in Antarctica, which means that all the voters fully understand what Antarctica is all about.

If you have the United Nations running it, you will have all sorts of little nations from all over the world involved who don't even know where Antarctica is, let alone have any knowledge of the environmental issues.

So I am expecting this to happen fairly soon. For the first forty years very little happened because the nations involved were too busy arguing the procedure about how to set up the Treaty.

Only quite recently, they decided to establish a formal secretariat, so there has been a secretariat for the Treaty set up in Buenos Aires.

Obviously the next move is for them to draw a

constitution, forming the government of Antarctica, by the Treaty Nations. I think within twenty or thirty years we'll see some formal international agreement for the Treaty nations to control Antarctica.

Given that might happen, what do you see as the future of mining in Antarctica?

I firmly believe there is no future for mining the continental shelf. I personally think that all the hoo-ha about mining in Antarctica is nonsense.

But mining off the coast of Antarctica is a different business. I am sure there is a danger of oil being drilled in the oceans around the coast of Antarctica.

I am not worried about anyone trying to drill for oil or to try to prospect for minerals on the continent. There is hardly any exposed rock and almost no ports, where ships could load ore. There's no point in worrying about drilling through the ice, because there is a mile of ice to get through and you are not going to prospect and find anything much.

When you look at what they have done already, geologists have crawled over every bit of exposed rock in Antarctica and they haven't found a single ore body, let alone one that has commercial value. So what with there being no apparent ore bodies, no port to service them and no fuel except what you take down for the machinery, anything on the continent is just not on.

Getting back to another area of your life Phil, you have always been a very motivated and highly focussed individual. That probably started from a very early age

197

at school from what I gather. You must have had some key mentors in your life. Can you talk about that?

First of all, I'll mention a guiding principle that I developed when I was quite young and one which I have had all my life. That is to **appreciate the value of time and that every second of your life is important**.

First of all, you mustn't waste time. I quite early decided that every year I'd do something which was worth doing. I would want to add foundations for my career: do some further study; or improve my qualifications; or learn to play an instrument; or learn a new language; or visit new places; or have new adventures.

At the end of each year, I wanted to look back and find the year had not been wasted and that something valuable had happened.

All my life I have followed that principle and I have lectured on it to university students, to convince them to make sure they don't waste their time.

The mentors in my life? The first one was my elder brother. I was the cautious one, very careful, very safety-first in attitude, never doing something unless I was sure it would work.

I had an elder brother, who was adventurous and quite haphazard. He was my brother Geof. He first of all broke the ties of the family constraints, by defying my parents and going and doing things in the mountains, when he was quite young.

That opened the door for me, because by the time I came along, my parents were used to him doing it, so there was no point in worrying about me. He and I had numerous adventures in the Grampians, where we spent most of our school holidays, because we lived nearby in Hamilton.

We had all sorts of adventures. We developed rock climbing skills, mountaineering and bushwalking, camping and so on.

Then we discovered that there was a place called the Victorian Alps which had snow. Of course we'd read about skiing in Antarctica. So we decided to get into the Victorian Alps.

We made skis and rucksacks and to afford skiing we never paid to stay in a chalet. Instead we stayed in cattlemen's huts or camped out. So that my ski experience then formed the background to my later work in Antarctica.

Long before I went to Antarctica, I learnt all about snow and ice, survival and skiing. I'd read all the polar books about Scott and Shackleton and all those other explorers.

I think I was more highly prepared for an Antarctic career than anyone in history except Amundsen; Scott, Shackleton and Mawson even. Mawson did a lot of fieldwork as a surveyor and geologist but not in any snow country. He had never been skiing, I don't think he'd ever seen snow or ice before he went to Antarctica.

Did you do much skiing in Antarctica?

No. There is very little chance to do any because around the coast there is very little drift snow. If you go inland to get a drift to ski on, it's too dangerous to take the risk.

Fair enough. I tried once and never again. It scared the hell out of me!

Around Mawson, which is a desert area, most of the surfaces are blue ice except in winter when you don't go out anyway.

So he was your first mentor. What about the next one?

The next was my Professor, Leslie Martin, Professor of Physics. He was not a professor when I did my first degree. I did my Master's under the previous Professor, Sir Thomas Laby, a very famous physicist. He helped me in a small way, because he showed me what a first-class, world physicist he was and how he operated. He retired while I was working at the University.

His place was taken by Leslie Martin who was his deputy in the Physics Department. He became a knight ultimately. The Commonwealth Government had made him its Chief Scientific Advisor. He was responsible for my entering Antarctic work. I have told the story earlier so I won't repeat it, about walking down the passage saying he was looking for a Chief Scientist.

So I had a long-term relationship with Les Martin, his wife and family. I also became very close friends with his son Ray, who became Vice-Chancellor of Monash University and who was a tennis player. He played tennis with me on numerous occasions.

The next thing with Les Martin, was that Melbourne University decided to set up a Faculty of Applied Science. Les Martin and Professor Greenwood, a metallurgist, were the two who dreamt up the idea. When they finally pushed it through Council, it created the Faculty of Applied Science.

They then decided, since I was about to be given an

honorary degree by Melbourne, they would use my fame at the time to publicise the new faculty.

My second honorary doctorate was awarded by the Faculty of Applied Science, which suited me fine, because the colour facade of the gown was very glamorous.

What year was that Phil?

Oh 1960 sometime. Incidentally, I recently noticed in the magazine issued by the Department of Education at Melbourne, they had just set up the Sir Leslie Martin Institute, to help produce better scholarship in Tertiary Education in Institutes.

I was going to tell you that Les Martin, having organised my honorary doctorate and having introduced me to Antarctica, was living in Camberwell in a very nice apartment on the top floor of a gracious home. He had decided to go and build a house of his own and he offered Nel and me his apartment. That was another point of his value.

He became chairman of the committee, set up by the Commonwealth, to report on the structure of Tertiary Education. It recommended that institutes of technology be set up, consequently Colleges of Advanced Education developed all around Australia. My getting into the VIC was a direct result of his committee's report.

Then, when he was Chairman of the Universities Commission, I had further dealings when, for a while, I was the Buildings Officer on the LaTrobe University Council. On one occasion when he came to look at the LaTrobe budget, I had to deal with him on the buildings budget.

So my path and his crossed intermittently over a number of years and was extenuated by my friendship with Ray Martin, his son.

Regarding other sponsors, it's hard to think of any particular one. Douglas Mawson was useful because he was on my Antarctic Planning Committee and he was put there because of his past association with Antarctica.

He was very supportive of what I was doing. His support for my plans was really valuable in the Committee, so that you could say that I traded on his support to push through various ideas that I had for Antarctica.

A lot of those people that you've talked about were 'Sir', but you were never given a Knighthood. Why not?

On the first occasion that I was offered an honour, my Minister, Lord Casey, put me up for a Commander of the Order of the British Empire (CBE). When it went to the Prime Minister, Robert Menzies, he knocked it back to an OBE.

I declined to accept the OBE because I said that all my counterparts, in other countries, had received their countries' highest honours. A CBE would have been acceptable, but not an OBE because if I got an OBE, all my counterparts would've thought that I had blotted my copy book.

So I wrote to the Honours Secretariat explaining why I declined and they tended to agree. It was many years later that I was made an Officer in the Order of Australia. Many years after that I was made a Companion.

Now I am not at all worried about having missed the Knighthood because the Companion is much higher than the Knighthood anyway.

202

The Order of Australia Officer, the AO, is just barely below the Knighthood, but the AC, the Companion, is very much higher.

That is the highest accolade that this country can give you.

Yes. It is higher than the Knight Commander of St Michael and St George (KCMG) for example, which is a pretty high Imperial Order. Of course the British knighthoods and awards have all disappeared now in Australia. There are no more Polar Medallists, either in Australia, for the same reason.

So Phil, if you were giving advice to young people today not to waste time, there must be other things that you feel led to your enormous success in life?

Two other things I'll mention. One is, **don't be frightened to jump at an opportunity**. Too many people are too cautious.

I had a friend whom I offered to send down to Antarctica for a year with the French to get experience. Knowing he'd be away for a year, he felt his job wouldn't be there when he got back.

He should have taken it because with that experience he would easily have found another job. But he was too afraid to jump. He's regretted it all his life. I still meet him and he always says, 'I should have taken it!'.

So my advice to young people is that if there is something exciting and looks reasonably good don't be afraid to jump at it.

For example, when I accepted the Antarctic job, all my mates said, 'You're stupid to accept a job which has no definite future and give up a permanent job at Melbourne University to take it.'

I believed the Antarctic had a future and I was prepared to take a risk, so I jumped. In a sense I jumped later to go to the VIC.

The other reason is, I believe that **it is a good idea to change jobs every ten or fifteen or twenty years**. If you stay too long in one job, it becomes too easy for you and you're not being stretched to your fullest capabilities.

When I jumped out of the Antarctic Division, I knew it all so well I did not have to refer to any files for anything. I found that I was just leaning back on the oars a bit, taking it easy, just drifting along. I felt if I got a new job, it would be a challenge, which it was. I was re-invigorated you might say. I found myself right back in a bigger, newfound career, like I had when I set up the Antarctic Division.

Ever since childhood you have been able to take a new challenge head-on and relish it. Too many young people won't do that which, appears a real problem.

In a sense there are all sorts of restrictions, one of which is at Government level. In a way I used to have carte blanche in Antarctica. The people in Canberra put very few constraints on me. They let me be the knowledgeable one in the system and they used to give consent in the broad, but never in detail.

Now, the Director of the Antarctic Division has to keep referring to Canberra on all sorts of minor matters. On the other hand he has much more money than I had and much more backup with staff, buildings, and logistic support. You can't have it both ways.

I guess you have been known as 'Lucky Phil'. Do you really believe you were lucky or did you just take advantage of the situation?

I think I deliberately created exploration avenues, which was the adventurous side of my life. I used to direct where to go and what to do. I used to arrange the logistic support, pick the men for the job, and they would carry it out.

In some cases, they would dream up ideas and toss them back to me and I would fully support any good adventurous ideas my men thought up.

There is one glorious example. We sent a team down to Casey and the Officer-in-Charge decided it'd be fun, and adventurous and something new, to run a tractor train into the Russian station at Vostok, nine hundred miles inland.

He reported back and asked me if they could do it. This was not my idea. This was the idea of the people at Casey. I considered it and I thought 'Yes, that would be great!'.

But they said they would need support from the Americans, because they would need to refuel the tractors. They tossed that back to me and I made the arrangements with the Americans to drop fuel to them from aircraft. The Americans were able to provide this logistic support for the adventure and consequently, the Casey crew carried out what became an amazing adventure.

They drove two tractor trains, led by a Weasel, nine hundred miles inland and nine hundred miles back, which was a remarkable feat for novices. They had never done anything like that before.

205

What year was that Phil?

I think it was about mid-1960s, I forget the exact date.

One of the beauties about the Antarctic Treaty is that everyone co-operates down there irrespective of politics. They always have and I guess they always will.

Is there anything else that you would like to add? For example, how many times did you think your number was up in Antarctica?

My book *You have to be lucky* describes these occasions. The worst was when we couldn't find the ship from the air. We had no direction finder. We had radio contact but fuel was running out and we couldn't find it. We tried various strategies, but none of them worked.

Finally by radio, I said 'Get every pair of binoculars in the ship and put the men on top of the bridge and divide the sky into segments and have each man looking at one sector of the sky to see if they can see our plane'.

One of the blokes with binoculars found a spot in the sky and they all focussed and saw it was an aeroplane. Over the radio they gave directions as to how to find the ship.

Our float plane had taken off from an open bay, about a mile wide. We had thought 'You can't lose that'. However while we were away, the pack ice had closed up. We had no place to land, because you can't land a float plane on an icefloe.

When we arrived back, the captain put the ship on full steam ahead. This meant it was only surging at a couple of miles an hour, but it was pushing the ice and the propellers were churning the ice madly to clear a lane behind.

206

Eventually we had a fifty-yard stretch of clear water which enabled us to land, with only a few gallons of fuel left.

If we'd have landed without being sighted we would never have been found. They would not have known to go north, south, east or west. If we had landed on an icefloe, we'd have crashed because the undercarriage would have collapsed. It was pretty close.

Was it a single engine aircraft?

Yes. All our flying was hazardous because of the single engine and the Beavers were single engine. Later on people used the Otter because they were twin engined.

To wrap up, what would be the advice you would give adventurous spirits?

People used to ask me would I have gone into space had I been offered the chance and I said, 'Yes! I'd have loved to have been a spaceman'. You have to have an adventurous spirit. As various opportunities open up you, decide whether or not they are worth doing. But I decided years ago, after lecturing and teaching, that only about ten percent of people are really adventurous. All the others are very cautious.

I think that is a fact of life! Thank you Dr Law.

APPENDIX ONE

VOYAGES

THE FOLLOWING IS A LIST OF VOYAGES
UNDERTAKEN BY DR LAW
TO THE ANTARCTIC REGION

December 1947 - March 1948: ANARE expedition to Antarctica and Macquarie Island in HMAS *Wyatt Earp* conducting experiments on latitude variation of cosmic rays

8 February - 1 April 1948: Senior Scientific Officer, Voyage in HMAS *Wyatt Earp* to Balleny Islands, Borradaile Islands, Commonwealth Bay and Macquarie Island

July-September 1948: Voyage to Japan in MV *Duntroon* by courtesy of the Australian Army to extend cosmic rays latitude variation measurements across the equator and tropics

21 January - 2 February 1949: Voyage in HMAS *Labuan* to Heard Island and Kerguelen Islands

January-March 1950: Australian observer, voyage in MV *Norsel* with Norwegian-British-Swedish Antarctic Expedition, Dronning (Queen) Maud Land

16 January - 1 March 1951: Voyage in HMAS *Labuan* to Heard Island and Kerguelen Islands

28 April - 19 May 1951: Voyage in SS *River Fitzroy* to Macquarie Island

9 February - 19 March 1952: Voyage in MV *Tottan* to Heard Island and Kerguelen Islands

24 March - 16 April 1952: Voyage in MV *Tottan* to Macquarie Island

12-27 December 1953: Voyage in MV *Kista Dan* to Macquarie Island

4 January - 31 March 1954: Voyage of MV *Kista Dan* to Heard Island, Kerguelen Islands and exploring in Mac Robertson Land

7 January - 23 March 1955: Voyage in MV *Kista Dan* to Heard Island Magnetic Island, Mawson and Kerguelen Islands

27 December 1955 - 26 March 1956: Voyage in MV *Kista Dan* to Lewis Island, Balleny Islands, Mirny, Mawson, Heard Island and Kerguelen Islands

17 December 1956 - 12 March 1957: Voyage in MV *Kista Dan* to Vestfold Hills, Davis, Mawson and Kerguelen Islands

7-28 December 1957: Voyage in MV *Thala Dan* to Macquarie Island

3 January - 19 March 1958: Voyage in MV *Thala Dan* to Lewis Island, Dumont d'Urville, Mirny, Davis, Mawson, Heard Island and Kerguelen Islands

6 January - 5 March 1959: Voyage in MV *Magga Dan* to Lewis Island, Dumont d'Urville, Wilkes, Oates Land and Macquarie Island

5 January - 11 March 1960: Voyage in MV *Magga Dan* to Dumont d'Urville, Davis, Wilkes, Lewis Island and Macquarie Island

29 November - 16 December 1960: Voyage in MV *Magga Dan* to Macquarie Island

22 December 1960 - 22 January 1961: Voyage in MV *Magga Dan* to Wilkes

24 January - 19 March 1961: Voyage in MV *Magga Dan* to Mawson, Dumont d'Urville, Oates Land and Macquarie Island

1-18 December 1961: Voyage in MV *Thala Dan* to Macquarie Island

22 December 1961 - 8 March 1962: Voyage in MV *Thala Dan* to Lewis Island, Wilkes, Commonwealth Bay, Dumont d'Urville, Oates Land and Macquarie Island

9 January - 24 March 1963: Voyage in MV *Nella Dan* to Heard Island, Mawson, Davis, Heard Island and Kerguelen Islands

2-17 December 1964: Voyage in MV *Nella Dan* to Macquarie Island

22 December 1964 - 15 March 1965: Voyage in MV *Nella Dan* to Mawson and Davis for extensive exploration

29 December 1965 - 11 March 1966: Voyage in MV *Nella Dan* to Wilkes, Mawson, Davis and Lewis Island

DECORATIONS, AWARDS, ACCOLADES AND MEMBERSHIPS

DR LAW HAD ACHIEVED WIDE RECOGNITION FOR
HIS CONTRIBUTION TO SCIENCE, EDUCATION AND
COMMUNITY LIFE.
HIS LIST OF MEMBERSHIPS SHOWS JUST HOW
WIDE THIS CONTRIBUTION HAS BEEN

Queen's Coronation Medal, 1953; Commander of the Order of the British Empire (CBE), 1961; Polar Medal, 1969; Officer of the Order of Australia (AO), 1975; Queen Elizabeth II Jubilee Medal, 1977; Companion of the Order Australia (AC), 1995

Award of Merit (Royal Life Saving Society), 1929; Award of Merit (Commonwealth Professional Officers' Association), 1956; Clive Lord Memorial Medal (Royal Society of Tasmania), 1958; Founder's Gold Medal (Royal Geographical Society, London), 1960; John Lewis Gold Medal (Royal Geographical Society of Australia, South Australia Branch), 1962; Two Thousand Men of Achievement Award, 1969; Vocational Service Award (Rotary Club of Melbourne), 1970; Men of Achievement Award, 1973; Sir Edmund Herring Memorial Award for Outstanding Service to the Youth of Victoria, 1982; Statue of Victory Award (World Culture Prize for Letters, Arts and Science), 1985; Gold Medal, Adventurer of the Year (Australian Geographic Society), 1988; James Cook Gold Medal (The Royal Society of New South Wales), 1987; National Award for Lifetime Contribution to Science and Technology (Clunies Ross Foundation), 2001

Matriculated, University of Melbourne, 1932; Bachelor of Science (University of Melbourne), 1939; Master of Science (University of Melbourne), 1941; Doctor of Applied Science, *Honoris Causa* (University of Melbourne), 1962; Doctor of Science, *Honoris Causa* (LaTrobe University), 1975; Doctor of Education, *Honoris Causa* (Victoria Institute of Colleges), 1978; Doctor of Applied Science, *Honoris Causa* (RMIT), 2008

Member, Ski Club of Victoria, 1936-1937; President, Boxing Club, University of Melbourne, 1943-1948; Chairman, ANARE Planning Committee, 1945-1965; Member, The Royal Society of Victoria, 1946- ; Member, Alpine Club of Victoria, 1947-1975; Fellow, Australian Institute of Physics, 1948; Fellow, Royal Geographical Society, London, 1949; Member, Antarctic Club, London, 1949; Fellow, Institute of Physics, 1950; Chairman, Victorian Division, Institute of Physics, 1951-1952; Chairman, Australian Committee on Antarctic Names, 1952-1974, 1976-1981; Member, Australian Committee, International Geophysical Year, 1953-1957; Appointment as Justice of the Peace, Australian Antarctic Territory, 1955; President, Geographical Society of New South Wales, 1955-1956; Member, Council, University of Melbourne, 1959-1978; Patron, British Schools Exploring

Society, 1960; Deputy-Chairman/ Chairman, World Health Organisation Conference on Medicine and Public Health in the Arctic and Antarctic; Member, Victorian State Committee, Duke of Edinburgh's Award Scheme, 1963-1980; Member, Recreational Grounds Committee, University of Melbourne, 1963-1988; Chairman, Working Group of the Committee of Investigation into Administration of the University of Melbourne/ Member, Committee of Investigation, 1964-1966; Member, Council, LaTrobe University, 1964-1974; President, Geelong Area, Victorian Scouts Association, 1964-1995; Executive Vice-President, Victoria Institute of Colleges, 1966-1977; Chairman, Australian National Committee for Antarctic Research, 1966-1980; Member, Council, Victorian Institute of Marine Sciences, 1966-1980; President, The Royal Society of Victoria, 1967-1969; Member, Advisory Council on Tertiary Education, Victoria, 1968-1977; Member, Board, Apex Foundation for Research into Mental Retardation, 1968-1978; Councillor, Science Museum of Victoria, 1968-1983; Trustee, Science Museum of Victoria, 1968-1983; President, Melbourne Film Festival, 1969-1970; Member, Victorian Universities and Schools Examination Board, 1969-1972; Chairman, The Royal Society of Victoria Committee to establish an Institute of Marine Sciences, 1969-1977; Member, Australian Council on Awards in Advanced Education, 1971-1977; Member, Australian Territories Accreditation Committee for Advanced Education, 1971-1977; President, Graduate Union, University of Melbourne, 1971-1977; Member, Committee for Natural Sciences, Australian National Committee for UNESCO, 1972-1977; Member, Victorian State Council for Technical Education, 1972-1977; Vice-

President, Sports Union, University of Melbourne, 1972-1979; President, Melbourne Film Society, 1972-1992; Trustee, Specific Learning Difficulties Association of Australia, 1972- ; Member, Science and Industry Forum, Australian Academy of Science, 1973-1982; Foundation Fellow, Australian Academy of Technological Sciences and Engineering, 1975; Member, Commonwealth Committee of Enquiry into Scientific and Technological Information Systems, 1975-1977; Chairman, Organising Committee, ANZAAS, 1976-1977; Vice-President, Australian Club of Rome, 1976-1979; Foundation President, Australian and New Zealand Scientific Exploration Society, 1976-1982; Freeman, Victorian College of the Arts, 1977; Hon. Fellow, RMIT, 1977; President, Victorian Institute of Marine Sciences, 1977-1980; Member, Australia and New Zealand Schools Exploring Society, 1977-1982; Fellow, Australian Academy of Science, 1978; Foundation President, Victorian Institute of Marine Sciences, 1978-1980; Deputy President, Science Museum of Victoria, 1979-1982; Fellow, ANZAAS, 1981; Hon. Life Membership, Melbourne University Sports Union, 1982; Chairman, Australian Scientific Exploration Foundation, 1982-1983; Patron, British Schools Exploring Society, 1983; Member, Antarctic Names and Medals Committee, 1984-1993; Hon. Life Fellowship, Museum of Victoria, 1985; Hon. Life Membership, Melbourne University Graduate Union, 1987; Trustee, RMIT Foundation, 1994; Foundation Fellow, The Royal Society of Victoria, 1995; Chairman, RMIT Foundation, 1995-1998; Member, Board of Directors, Glacier Society, USA, 2001; Chancellor, The Royal Societies of Australia, 2008

Member: ANARE Club (Founder); The Kelvin Club; Melbourne Club; Melbourne Cricket Club; Royal South Yarra Lawn Tennis Club, The Antarctic Club

APPENDIX THREE

PUBLICATIONS

THE FOLLOWING IS A LIST OF DR LAW'S
SIGNIFICANT PUBLICATIONS,
AND THOSE ABOUT HIM

Report on tropic proofing of optical instruments in New Guinea (Melbourne: Scientific Instrument and Optical Panel of the Ministry of Munitions, 1944)

Nutrition in the Antarctic (Melbourne: Royal Australian College of Physicians, 1957)

The Antarctic voyage of MV Thala Dan, 1958 (London: The Geographical Journal, 1959)

The Vestfold Hills (Melbourne: Antarctic Division, Department of External Affairs, 1959)

Australia and the Antarctic - John Murtagh Macrossan Memorial Lectures, 1960 (St Lucia, Qld: University of Queensland Press, 1962)

Resources of Australian Antarctica (Melbourne: 1963)

Antarctica 1984 - Sir John Morris Memorial Lecture, 1964 (Hobart: Adult Education Board of Tasmania, 1964)

The exploration of Oates Land, Antarctica (Melbourne: Antarctic Division, Department of External Affairs, 1964)

Changing pattern of requirements in professional education (Melbourne: College of Nursing, 1967)

Planning for uncertainty: a conference on the planning facilities for tertiary education (Canberra: Australian National University, Centre for Continuing Education, 1973)

Antarctic Odyssey (Melbourne: Heinemann Australia, 1983)

The Antarctic voyage of the HMAS Wyatt Earp (St Leonards, NSW: Allen & Unwin, 1995)

You have to be lucky: Antarctic and other adventures (Kenthurst, NSW: Kangaroo Press, 1995)

Report on the voyage of M. V. Kista Dan, *January - March 1955* (Melbourne: 1955)

Papers of Phillip Law, 1870-1999 [manuscript] (19??)

jointly, with John Bechervaise

ANARE: Australia's Antarctic outposts (Melbourne: Oxford University Press, 1957)

Lennard Bickel, interviewer

Interview with Phillip Garth Law, scientist, administrator and Antarctic explorer [sound recording] (1975)

Oscar Manhal and Gavan McCarthy, compilers

The records of Phillip Garth Law (1912-): deposited with the National Library of Australia and held by PG Law (Parkville, Vic: Australian Science Archives Project, 1989)

Janet L Mentha and Graeme F Watson, editors

Education, Antarctica, marine science and Australia's future: proceedings of the Phillip Law 80th birthday symposium, 29 April 1992 (Melbourne: The Royal Society of Victoria, 1992)

Roy Norry, illustrated by Don Angus
Antarctic explorer: the story of Dr Phillip Law
(Melbourne: Nelson, 1993)

Kathleen Ralston
A man for Antarctica: the early life of Phillip Law
(South Melbourne: Hyland House, 1993)
Phillip Law: the Antarctic exploration years, 1954-66
(Canberra: Ausinfo, 1998)
Antarctic leader and administrator: the early life of PG Law [microfilm] (Clayton, Vic: Monash University, 1992)

Australian Antarctic research expedition, 1940-50
(19??)

Frank Heimans
Phillip Law, scientist and explorer [videorecording]
(Lindfield, NSW: Film Australia, 1993)

Ian G Toohill, author and photographer and Phillip G Law, narrator
Mawson Base - Face to Face (Melbourne, Curriculum Branch Education Department, 1983) [videorecording]

Australian National Antarctic Research Expeditions, 1947-1966 [cartographic material]: under the direction of Stuart Campbell, 1947-48, and Phillip Law, 1949-66 (Canberra: Australian Surveying and Land Information Group, Department of Administrative Services, in collaboration with Antarctic Division, Department of the Arts, Sports, The Environment, Tourism and Territories, 1989)

218

For a full list of Dr Law's publications, please refer to the following website: http://www.austehc.unimelb.edu. au/guides/lawp/LAWP.htm

IAN TOOHILL

BORN IN Melbourne in 1949, Ian grew up and completed his education in the northern suburbs. Graduating as a primary teacher in 1970 he started his teaching career in 1971 and then began a Graduate Diploma of Fine Art at RMIT, attending the course at night majoring in sculpture. He completed his diploma with twin majors in Sculpture and Photography. After five years in primary schools, he spent two years as a state wide curriculum arts advisor with the Victorian Education Department and attended LaTrobe University where he completed a Bachelor of Education majoring in Media Studies. Ian then returned to teaching as a primary school assistant principal for three years. In 1981 he took up a position as a curriculum materials producer for the Education Department and spent six years in that role.

At that point he decided it was time for a 'sea change' and took family leave from his teaching career to run his own business. Eight years of hard labour ensued running a family based photographic and video production company.

In 1994 Ian returned to teaching in a Melbourne secondary college as a specialist in Media studies and photography, a position which he still holds today.

Concurrently with his long career in education, Ian had joined the Army Reserve (then known as the CMF) at the age of eighteen and subsequently completed his National Service obligation. He is still a serving officer having attained the rank of Major. After spending thirty years as a logistics officer he transferred to Defence Public Affairs in 1998.

It was while posted to Melbourne Water Transport Unit that he was selected as a member of the Army Detachment to ANARE and was deployed to Antarctica for the summer of 1982-83 in ship to shore logistic support. Ian's forty years of military service have seen him involved in Defence Support Operations, both at home and abroad.

He has made a life long commitment to community service and, as an active member of Rotary international for twenty years, has been involved with in-country assistance to Fiji, Nepal and East Timor in particular.

Ian is somewhat of an adventurer himself with a passion for the outdoors and challenging projects. He has trekked and climbed in the Himalayas, navigated across Lake Eyre (in flood), is a keen sailor and enjoys outback four wheel driving. He is also a musician and currently plays in a revival rock band when his busy life permits.

During his long and varied career Ian has produced a range of print, film and video resources as well as photographic exhibitions for himself, private clients, the Australian Defence Force and Victoria's Education Department.

INDEX